Building PDA Databases for Wireless and Mobile Development

Building PDA Databases for Wireless and Mobile Development

Robert Laberge
Srdjan Vujosevic

Wiley Publishing, Inc.

Publisher: Robert Ipsen
Editor: Carol Long
Developmental Editor: Adaobi Obi Tulton
Managing Editor: Fred Bernardi
New Media Editor: Brian Snapp
Text Design & Composition: Benchmark Productions, Inc.

This book is printed on acid-free paper. ∞

Published by Wiley Publishing, Inc., Indianapolis, Indiana

Published simultaneously in Canada

Library of Congress Cataloging-in-Publication Data:

Laberge, Robert, 1961–
  Building PDA databases for wireless and mobile development / Robert Laberge, Srdjan Vujosevic.
    p. cm.
"Wiley Computer Publishing."
Includes bibliographical references and index.
  ISBN 0-471-21645-3
  1. Mobile computing. 2. Database design. 3. Portable computers--Programming.
I. Vujosevic, Srdjan, 1961- II. Title.
  QA76.59 .L33 2002
  005.265--dc21

2002012145

Printed in the United States of America

10 9 8 7 6 5 4 3 2 1

Thanks to all the many firms and individuals who partnered with us in the creation of this project

—Robert Laberge

I am dedicating this book to all the good and professional people at BDP, Business Data Services, Toronto

—Srdjan Vujosevic

CONTENTS

ABOUT THE AUTHORS

Robert Laberge

Bob has been involved in the IT industry for over 20 years. Beginning in Montreal, he worked his way across Canada and down along the West Coast of the United States. He started as a mainframe programmer and soon became a curious database administrator, which led to many interesting insights and contracts. He currently concentrates on data warehousing, systems integration, and wireless systems.

Bob is mostly self-taught, as he believes there is much more available from books and manuals than from educational institution curriculums in the field of computer science—or at least it used to be that way!

Mr. Laberge is a co-founder of WaveDev, a wireless Internet venture, and a co-founder of Worldjobmart.com, a global Internet job site. He has had articles published in Microsoft's *MSDN*, *Web Techniques*, and several other popular magazines.

Srdjan Vujosevic

Srdjan has over 17 years of IT experience. Originally from Yugoslavia, Mr. Vujosevic is familiar with a multitude of IT technical issues and related business concerns.

Initially trained as an electrical engineer, Mr. Vujosevic holds intimate knowledge of computer internals (PC, Unix, and Mac), networks, WAP, and many programming languages. In addition, he has many years of database administration experience, which led him to many successful system integration projects.

Mr. Vujosevic is a co-founder of WaveDev, a wireless Internet venture, and a co-founder of Worldjobmart.com, a global Internet job site. He too has had articles published in Microsoft's *MSDN*, *Web Techniques*, and several other popular magazines.

An Introduction to Wireless and Mobile Applications

Initially, wireless mobile applications, for small portable devices, were programs totally unconnected from corporate or personal computer systems. That is from a real-time point of view. The majority of these types of applications were on Palm Pilots and were limited to address books, personal schedules, to-do lists, and so on. That's great for the traveling salesperson or busy corporate consultants, but not that useful in everyday life for most of us.

Then came the wireless Internet. This was thought to be the best thing since sliced bread in the computer industry—and is quite amazing. However, there are limits to business and everyday usage primarily because of the available devices. These wireless portable devices and special Wireless Application Protocol (WAP) cellular telephones are very limited in memory and display screen size, and are also difficult to enter information into because of the awkwardness of typing on the units themselves. Nevertheless, the technology is excellent. The ability to connect to the World Wide Web from nearly anywhere at anytime is truly a major advancement in the current borderless Information Age. For more information on WAP, including detailed tutorials and programs on how to build WAP applications, refer to our previous book, *WAP Integration: A Professional Developer's Guide*, by Srdjan Vujosevic and Robert Laberge, also published by John Wiley & Sons.

With the wireless Internet, we no longer have to be seated in front of a desktop personal computer physically hardwired to personal or corporate networks to reach our applications and databases. With portable wireless mobile technology, we can easily connect to the Internet or intranet and onto our legacy systems to obtain the most valuable asset of most businesses today: the stored data. This data is usually specifically structured and organized into related entities and objects known to the business users as information. We've all heard the expression "information is power," and to empower our workers with all the information they require when they require it wherever they require it is an absolute advantage to any business.

To sum it up, portable wireless mobile computing is the ability to retrieve information from the main computer system and/or repository of information on a wireless device at anytime and anywhere. An obvious plus here is the ability to also enter or update information on those main computer systems remotely via our wireless device. This gives us all the freedom to theoretically manipulate all information from afar.

The specific type of wireless devices used in our portable wireless mobile architecture is the *Personal Digital Assistant*, better known as the PDA. The usage of these types of devices is growing at an alarming rate. In the next five years, most cellular telephones will have Internet capabilities, and a large number of individuals will be porting a PDA of some sort.

Palm Inc. used to have nearly the entire market with their Palm Pilot product line, but now Microsoft has jumped into the arena with their own devices called Pocket PCs. There are other types of PDAs, but we'll be focusing on Palm devices, since they've been around for a long time, and Pocket PCs, since we believe that these will dominate the marketplace in several years.

These PDAs are fairly small, but larger than most cellular telephones. They contain different operating systems depending on who built the unit, and have a multitude of programs. They hold plenty of information and are generally self-contained. The future of these PDAs is to simulate mini-laptops with operating systems paralleling current desktop personal computers.

A major enhancement of these personal devices is and will continue to be the evolving speed and memory capabilities. As the memory increases, the capabilities of the units will also increase, allowing desktop applications onto PDAs—which will truly open the industry to portable mobile synchronized wireless applications. Imagine writing a note on a PDA while riding on a train or plane and having it load up on your main computer system halfway around the globe. Imagine downloading a 5GB database directly onto your PDA and drilling down into the data via a data mart front-end. The data warehouse extraction, transformation, and loading routines will one day completely involve PDA databases, perhaps as the data mart receiving agent and as another source system.

The portable wireless mobile architecture referred to previously is exactly this type of design. Wireless devices will become portals to enterprise systems and repositories of remotely gathered data, while holding a subset of the enterprise model. Current PDAs already have the capability to expand to 5GB of memory, meaning that they can hold entire application databases and the applications themselves.

This is exciting stuff for architects, programmers, and businesses. Imagine one day hosting a Web site with an Access, Sybase, DB2, or Oracle database on your Palm Pilot or Pocket PC!!

The technology to build applications with solid databases is here today. Wireless PDA memory can be expanded from the original 16, 32, or 64MB to a whopping 5GB. PDA processing power is already at 133 MHz up to 233 MHz, which is a tad below the CPU running the laptop on which we're writing this chapter. All the pieces are available today. With memory and CPU strength, the only thing missing is the ability to connect to the enterprise system. Well, surprise, surprise, connectivity is here, too, and improving every day.

Current network capabilities include WAP if the physical units have browsers and cellular connections, wireless modems, which connect directly to the Internet, wired connections to cellular telephones having WAP, and directly to the host computer. The last option is the most practical for any wireless PDA for full downloads or uploads, and we'll discuss in more detail later in the book.

There are many methods of connecting to the Internet, intranet, or a specific computer, and all are viable business solutions for today's large, medium, and small corporations. The price for the individual units and all expandable options won't break the bank either and is quite affordable to individuals.

With all the pieces available, network capabilities, affordable pricing, and the ability to design and develop applications for PDAs synchronizable with personal computers or host enterprise systems, portable mobile wireless systems and databases are certainly in our future.

About This Book

We believe the essence of this book is captured within its title: *Building PDA Databases*. The goal of this book is to introduce to you, the reader, how to build databases on personal digital devices to be used with PDA applications. Of course, we also try to show how to build PDA applications to use the data within the database.

Several years ago, the business community accepted the Internet as a viable solution to extending the corporate environment. Today, there is another movement toward enhancing this methodology, and it is via empowering mobile wireless business users. Employees who are regularly away from their offices and yet want to be up-to-date on their firm's information, inventories, offerings, and whatever other information can help them in their daily roles, want more access. Salespeople want to know pricing, inventories, delivery schedules, and so forth. We're no longer just looking for quick access to email, we're now asking for full access to applications and corporate enterprise-wide information.

With the popularity of mini-computers such as PDAs in the business environment, and the expanding processor speeds and disk (memory) capacity, we're seeing more requests for business information and applications to access that information. As with any new technology or availability of technology, after a short period of settling in, the application of that technology begins to emerge. Individuals and organizations begin to vision new methods of applying the technology, and this rolls over to real-life usage.

We believe that the time to apply this new technology is here, and we hope this book helps individuals and firms in viewing the possibilities of using PDAs and data from corporate environments in everyday scenarios.

Is This Book for You?

If you're holding it, it's for you unless you just like holding technical books. This book was written for programmers, technical managers, technical architects, and anyone wanting to learn about PDA database applications. If you want to learn the basics, this book will certainly give you that. If you require a reference or a how-to manual, again we hope this helps.

The idea of this book was to create a reference manual, and to include many working functional programs and applications for PDA (Palm and Pocket PC) database applications. With some programming experience and the examples in this book, along with the program code available on the site www.wiley .com/compbooks/laberge, you should be able to start programming within several days.

Covered Material

We try to cover details of each topic leading to databases on PDA devices, including an introduction to databases, client-server methodology, data warehousing, and so on. Once into database vendor offering chapters, we create many different applications for a how-to implementation. We also walk through most program code line by line, explaining all the pieces.

- Chapter 1, "Introduction to PDAs," is a basic introduction to the mobile wireless environment. It includes some basics, a quick view of WAP technology (the topic of our first book, *WAP Integration*, which is really Internet on cellular telephones), and an overview of PDAs and their environments.
- Chapter 2, "Database Architecture," discusses databases in general. It begins at zero and brings the reader step by step into what is a database.
- Chapter 3, "Client-Server Architecture," is an introduction to client-server methodology. This chapter describes what client-server is and how it applies to PDAs.
- Chapter 4, "Data Warehousing," deals with an advanced database topic called *data warehousing*. Many enterprises are very interested in centralizing their departmental information for everyone in the organization to access—this is data warehousing. We believe the accessing of this information will be very valuable to mobile employees, and therefore discuss some basic issues when applied to the PDA environment.
- Chapter 5, "Palm," brings the reader up to par on Palm devices. We present a very good overview of Palm as a whole, and of the products and operating system.
- Chapter 6, "Pocket PC," brings the reader up to par on Pocket PC from Microsoft. This chapter is a definite must if you are not familiar with devices using this technology.
- Chapter 7, "Mobile Application Development Tools," describes a handful of mobile application development tools. If you plan to access data on a PDA, you're probably thinking of building an application, and this chapter discusses several tools for this process.
- Chapter 8, "Palm's Database," discusses Palm's offering and how databases are deployed on Palm OS devices. Our example explains how to create an application to access data from the PDB database.
- Chapter 9, "Microsoft," introduces the Pocket PC offerings from Microsoft. This is a detailed chapter on the many aspects with an interesting application explained in detail.
- Chapter 10, "Sybase," introduces the Sybase solution to PDA databases. Again, this chapter discusses the components from Sybase, and has examples using specific pieces of the offering.
- Chapter 11, "IBM," introduces IBM's DB2 solution. A discussion on the components and an easy-to-follow application using the components follows.
- For your reference, we have a list of acronyms and appendices describing certain fine points from areas in the book.

The Web site (www.wiley.com/compbooks/laberge) has all the programs and applications described in this book, plus several more bits. Review them, copy them, and modify them. We created them for you to learn and use as templates for your own development efforts. We hope this book, its material, and associated programs will help you learn and develop your own PDA applications. Once you get the knack, you should be able to integrate your newly acquired knowledge within existing corporate environments, allowing the extension of your enterprise to a new wireless and mobile portal.

If you have any comments or questions, please feel free to email us at authors@wavedev.com. We would certainly enjoy your views and hearing about any specific applications you've built using these templates.

Enjoy.

CHAPTER 1

Introduction to PDAs

This chapter provides a brief history of the evolution of mobile wireless databases and introduces the Personal Digital Assistant, or PDA, and typical PDA environment. The purpose is to detail all the pieces of this technology at a high level. For those with no knowledge of PDAs, this chapter is an absolute necessity.

The Basics

The PDA environment and units can be very simple or quite advanced, depending on your level of technical expertise and usage. The term *PDA* is highly used, but it should really be *handheld computer* or *handheld PC*. PDA was initially used because the units mainly contained a daily calendar, personal address book, calculator, to-do list, perhaps a currency exchange program, and usually an international time zone map. In this respect, these units were indeed personal digital assistants, as people would forgo their usually big personal calendar and to-do agendas for these little electronic machines. I remember back in the early 1990s when I received my first PDA, a Texas Instruments digital assistant. I entered all the telephone numbers for all the people I knew, along with their addresses and whatever comments I could find such as birthdays, anniversaries, and upcoming special events. It was fantastic—the calendar would sound an alarm on the days I marked for notification. In meetings, I could take brief (very brief) notes and was a whiz with my fancy calculator that really couldn't do much more than just the basic functions. We're obviously still using the term *PDA*, as per its embedded use in the name of this book, but we really mean a handheld personal computer whenever we refer to it here.

A funny thing about the term *PDA* is that no vendor really uses it. When people are asked what a PDA is, they usually reply "Palm Pilot" (if they remember these), and recognize Pocket PC devices in the same category, but no vendor actually calls their devices "PDAs." "PDAs" was coined back in 1992 when introduced by Apple's Newton MessagePad, which really didn't take off. Microsoft also had their hands in these PDAs with WinPad, but had the same problem. Microsoft poured on the research and development and came up with quite a number of new product releases, including PC Companions, Windows CE, Handheld PCs, Palm-size PCs, Auto PCs, and Pocket PCs. Whenever we refer to PDAs in this book, we are referring to Palm devices such as the m500, and Pocket PC devices such as the Compaq iPaq.

As previously mentioned, the PDA's basic functions were quite simplistic. Palm's devices have expanded upon the PDA base with handwriting recognition, and enhanced memory—it used to be 64K at best, and now we're into megabytes and gigabytes. Along with these, PC synchronization, expansion slots for additional memory, and email have been added. Today's users can even buy added-value packages of memory with embedded programs for many uses. By far, the most popular added functionality is games. One of the latest Palm devices is the m500 model, which we'll be using throughout this book, compliments of Sybase, one of our main sponsors.

Palm opened their operating system, allowing thousands of developers the opportunity to create sophisticated and diverse programs specific to Palm. This really allowed the broadening of Palm's operating system, and hence device usage, that sent the PDA market skyrocketing. New units have color screens, modems, and the capability to wirelessly synchronize data and applications almost anytime and anywhere via modems and the wireless Internet—more on these possibilities later in the book. We'll show you how we built a simple application with its own database, and how we keep its data on the PDA unit and on the PC synchronized.

Pocket PCs are devices with the Windows CE operating system. Windows CE (originally code-named Pegasus in 1994) was initially based on Windows 3.1. Over the years, it has been completely rewritten as a 32-bit operating system built to run on embedded devices. WinCE basically manages the communication between the hardware and the applications that run on it. This operating system is very modular and therefore can run a multitude of different hardware platforms and applications. WinCE is used in cars, gasoline pumps, and video games, to name a few.

Pocket PCs are specific devices running Windows CE version 3.0, such as the Compaq iPaq 3800 series, which is what we'll be using throughout this book in our example applications (sponsored by Sybase). These devices also run Pocket Outlook, Pocket Office (Pocket Word, Pocket Excel), and much more. These devices truly are mini computers.

Portable Wireless Evolution

The wireless evolution began around 1996, but this depends on exactly what we're talking about. If we're discussing small computing devices such as Palm units or Pocket PC devices, these began in 1996 and 1999–2000, respectively. If we're talking about wireless connectivity to the Internet and hence connectability to corporate data via wireless devices, then we're talking 1997 with WAP.

What Is WAP?

The later scenario, called the *wireless Internet evolution*, used something called the *Wireless Application Protocol* (WAP). This technology came about only several years ago back in 1997. WAP is a method of global open wireless Internet standards for real-time communication of wireless mobile devices such as Web cellular phones, PDAs, and the Internet. WAP is not only a language, but also a platform for development and interconnectivity.

WAP components include:

- Multiple programming languages such as Handheld Device Markup Language (HDML) and Wireless Markup Language (WML).
- Physical units or wireless devices also known as *Web phones*, Web-enabled devices, or WAP devices built specifically with the capability to access the Internet. This additional built-in feature is called a *microbrowser*. Note that the primary purpose of the individual device, at this stage, is still to function as a telephone or PDA.
- Gateways that handle the transition of data from the wireless network through to the Internet network, and vice versa.
- Software Developer toolKits (SDKs) for the development and testing phase of the WAP applications.

WAP Evolution

For the first two years, no one really heard too much about WAP. In late 1999, however, something clicked and the world suddenly became very interested in wireless Internet technology. It could have been because a similar technology called *iMode* was making it big (really big) in Japan. Or, perhaps it was because the big cellular companies were getting their infrastructures set up and investing more funds in huge marketing campaigns. Either way, the technology allowing cellular phones to connect seemingly directly to the Internet was becoming very popular almost everywhere except North America.

As with any new technology, as WAP became more understood, acceptable, and feasible, people began thinking of potential revenue-generating opportunities, which led to research and development funding. With more financing available, more discoveries were made, more advancements were accomplished, and the wireless Internet evolution was under way. Some call it the Internet's second phase; others coin it as the merger of the Internet and telecommunications. Whatever you call it, it's definitely here and it's a major step in wireless mobile Internet communications.

WAP View

Mobility first came about from the telecommunications industry. It allowed individuals to communicate via wireless cellular phones, which gave us the ability to walk and talk, hence, the term *mobility*.

The Internet infrastructure has also been around for some time. This infrastructure allows anyone with a computer and a modem to connect to any other computer with a modem, and gives us the freedom to transfer or browse data from anywhere with connections to the Internet. However, unless you have a wireless modem, the communication must be wire based and, therefore, only *portable*.

Mobility and portability are now easily possible through the use of WAP, which is the merging of these two industries and technologies. Using wireless telecommunications and the Internet via wireless Internet devices, anyone can now access information anytime, anywhere.

With the world rushing toward a standardized telecommunication infrastructure, network, and protocol, along with the ever-increasing wireless device capacity, memory, and functionality, more people will be accessing the Internet via Web phones and PDAs than from any other method. While personal computers will always be our personal base stations at work or in our homes, wireless Internet technology will evolve into our everyday portable and mobile interconnectivity medium. The future potential is unlimited as it only has today's boundaries to overcome.

Growth Predictions

As seen in Figure 1.1, the forecasted number of cellular phone subscribers will increase enormously in the first half of the current decade due to several fac-

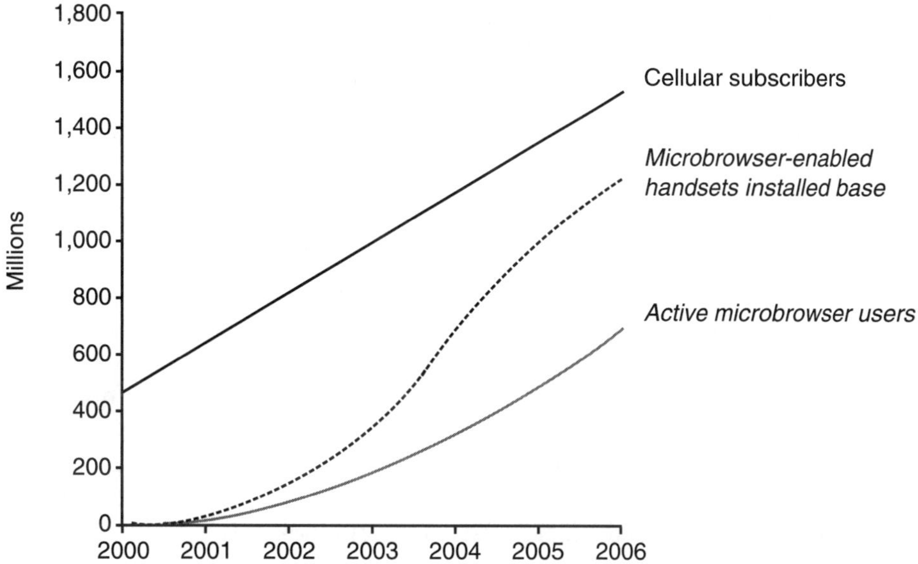

Figure 1.1 Wireless subscribers.
Source: Ovum (WAP/A)

tors. First, the ongoing trend of individuals wanting to be connected or available whenever they want, anytime, anywhere, is increasing. Second, many countries currently have poor telecommunication infrastructures, and cellular telecommunications is simpler and less expensive to set up.

Many people in third-world countries still do not have telephones, let alone cellular phones. Interestingly enough, telecommunications companies in these types of regions find it more cost efficient to simply erect cellular towers with relay base stations supplying wireless communications, rather than laying the groundwork to install cables throughout their cities. The cost difference between erecting several cellular towers and base stations compared to telephone poles and wires throughout thousands of miles of undeveloped rural regions is considerable. Based on these facts and the population growth of many world regions, the number of cellular subscribers will dramatically increase in the coming years.

With the knowledge and foresight of this new wireless Internet technology growth, wireless device manufacturers are beginning to distribute cell phones with microbrowsers in hopes that many individuals will subscribe to the service. Of course, there is a partnership between the microbrowser vendors, cell phone manufacturers, and the major cellular providers, since they will all profit from this new technology. The number of people who will subscribe to the wireless Internet service via their cellular phones will far surpass the number of individuals currently accessing the Internet via personal computers.

The applications available today are few, but we are in the infancy of the wireless Internet potential. In 10 years, many won't comprehend how we ever lived without wireless access to the Internet.

In the near future, we'll have the capability to contact others and have any information we desire at our fingertips. We'll carry our wireless mobile devices with us all day everywhere we go, and have unlimited access to any Internet item. The only foreseeable problem will be with wireless network and physical device capacities.

With so many individuals accessing the Internet, will the cellular carriers be able to cope with peak access periods as Internet service providers (ISPs) do now? Will the physical unit processors be able to cope with the mass influx of information at our fingertips? Look at the personal desktop computer; every year and a half, a bigger and better CPU chip is released. Instead of megabytes, PCs now come with gigabytes. As it stands, the Internet has only really taken off in the last half decade, and look at what's happened to the typical home and work personal computer within that time. Back in the early 1990s, it was sometimes difficult to justify the allocation of a 10MB database, and now a gigabyte single database is the norm. The processing power of some office computers today can easily replace the entire computer center of a typical company just 15 years ago. We've come a long way, and the same thing will happen with wireless devices, but at a much quicker pace. In 5 to 10 years, we'll have all the capabilities of our desktop computer on our cell phone or *wireless device*, as they'll be known.

NOTE

A while back, a news report on the radio mentioned a very interesting statistic: One out of five U.S. residents who wanted to learn more about the 2000 federal election between Gore and Bush obtained poll results and other information from the Internet, as compared to only 4 percent during the last election. That's an increase of Internet usage of over 600 percent in four years, which goes to show the amazing growth of this Information Age.

PDA Evolution

The PDA is the other half of the wireless evolution, in our context. WAP-enabled cellular phones are wonderful, but have only simple browsers to specific Internet sites. Rather than using Microsoft Explorer or Netscape Navigator from PCs, you would use a microbrowser from, say, Openwave or Nokia already pre-installed in your Web-enabled cellular phone. PDAs, on the other hand, are essentially tiny processing units with the capability to compute and synchronize features with a desktop personal computer.

PDAs evolved from personal organizers. Remember those expensive three- by eight-inch units, which opened into a screen and keyboard, or other units such as the one from Texas Instruments? These original devices are still active and still on the market. Their intention was to combine a calculator, calendar, personal telephone directory, a note pad, and maybe a currency exchange program. What a pleasure it was for people when they got their first TI personal organizer. They felt like they had a tiny super computer in their hands, with everything at the touch of a button wherever they were. That was technology!

The late 1980s and early 1990s gave us much technology, but the late 1990s were when the real revolution began. An initial player in the PDA business was Palm, who created and introduced the Pilot organizer back in 1996. Soon after came the famous PalmPilot, followed by the even better known Palm III series and Palm V series.

Along with these devices came Palm's HotSync software. We will be discussing this more in depth later, but at this point, it suffices to say that HotSync is the software used by Palm to connect the Palm device to the PC.

Palm currently holds approximately 60 percent of the world's PDA device market, and for quite a while they were alone in the industry. Then came Microsoft with their Pocket PC using Windows CE. With the PDA market anticipated to be very large and, very possibly, the future of computing, Microsoft couldn't resist and decided to bring on their own unique version of a PDA. WinCE is a small version of Windows used in many types of small devices, with the most common being the Pocket PC. It has also been used in many types of applications such as TV set-top boxes, factory-floor devices, cell phones, bar code readers, automatic teller machines, digital cameras, and so on. We, the authors, personally expect Pocket PC devices to continue to grow in number, models, and popularity. As with the Internet browser wars, the PDA battle has begun.

To parallel Palm's HotSync software, Pocket PC uses ActiveSync to connect devices to PCs. We'll look at this in the next section.

One of the most popular Pocket PC devices today is the Compaq iPaq, which is what we used to build the many Pocket PC applications in this book. Of course, this is just one model, but it's the one that currently dominates the market.

PDA Environment

This section highlights the basic hardware and connectivity software. Let's take it one step at a time and look at the components. The basic components are the physical units, the connectivity devices, and additions based on your requirements.

Devices

There are many types of devices; some are specific to the Palm, and others are specific to the Pocket PC world. Both devices connect to Windows on the PC, but Palm has no Windows components itself.

While the Palm m500 is an exceptional device, the iPaq has some very interesting features. It contains built-in Bluetooth software capable of wirelessly connecting to other Bluetooth devices. It also comes with a default voice recorder, which is very handy if you'd like to quickly record a memo or two.

Both types of devices have three components: the screen, the buttons, and the stylus—no mouse, no disk, no keyboard, and no cables (unless you count the cradles).

Screen

The screens work on the same concept of tap once to activate the program. The Palm unit has two parts, the display area and the Graffiti area, which also has the soft buttons. The display area shows the application, whatever that might be. The Graffiti area is used to write text, which in turn is stored in the application if applicable. The soft buttons in this part of the screen are predefined; that is, the little house is for home (main screen), the drop-down button is under that on the left side, the top right is to invoke the calculator, followed by the find button beneath it—nice and fast action buttons.

The Pocket PC is a bit different. Microsoft has arranged their screen similar to Windows 95 or 98. Top left is the Start button similar to a regular Windows PC. The main screen is called the Today window and contains many application summaries as configured by the user—the default is date/time, owner information, appointments/tasks, and email summary. The top bar has a little speaker icon for volume followed by the time. The bottom bar has the word *New*—click on it to add a variety of items. The bottom right icon is used for Bluetooth.

Buttons

Both devices have four basic buttons, which have quite different functionality on each device. The default buttons on the Palm, from left to right, are assigned to the calendar (or date book), address/phone book, to-do list, and memo pad function. The default iPaq buttons are, again from left to right: calendar, contacts, email inbox, and itask. Buttons on both devices can be reprogrammed to any application on the device. This allows for independent applications to be assigned to specific buttons. Both devices also have the middle-scrolling button (two buttons on the Palm unit). The iPaq has one more button on the left side of the unit. This button defaults to voice recording, as the unit is capable of recording sound.

Backlight

Both units have backlight screens. On the Pocket PC, the light comes on automatically when in use. After several minutes (depending on configuration), the unit's light will go off but the unit will remain on. For the Palm m500, simply hold the On button (upper right) for two seconds (also customizable) and the light will automatically appear. An obviously great feature to read the unit's screen in the dark, but it reduces battery life considerably.

Persistent State

Both units have a feature called *persistent state*. This means that when the unit is turned off and back on again, the last application running will return with all values as it was—if you were running something. This is similar to sleep mode on your PC. When your screen saver comes on, you simply move the mouse and your screen returns. The same applies to the PDAs—turn on the unit and the last thing you were working on returns. This feature is a must since it conserves battery power. By the way, a nice thing about these devices is that the battery lasts for 12 hours or more (depending on usage). A laptop battery normally has a short life of two hours, which can be frustrating when working away from an electrical outlet. The Palm m500 seems to last considerably longer than the iPaq does, battery wise, which might have to do with the backlight draining the life from its battery.

Reset

Both units have a Reset button; Palm's is on the backside top middle and easy to spot, since it says, "reset." iPaq's Reset button is located on the bottom near the cradle port. There is a soft and hard reset. Hard reset will completely reset the device, replacing all values back to the factory settings and removing any nonfactory applications. Beware: This will cause loss of data. Soft reset will simply stop and restart the unit.

On the Palm, press a bent paper clip or the special stylus device into the unit for a soft reset. For a hard reset, hold down the Power button and press the stylus reset device into the button opening. Check out the stylus; the top unscrews, revealing a smaller stylus specifically designed to reset the unit.

The Pocket PC unit follows the same rules, but the original stylus fits into the reset opening area. One push and the unit is stopped and restarted. To hard reset the Compaq iPaq, press and hold the two outside application buttons, insert the stylus into the reset switch area, and press the switch for five seconds. Then, to reactivate your unit, insert the stylus into the reset switch again and press it for one second—or simply connect the unit to the cradle. If ever you forget your device password, you'll have to perform a hard reset.

Connectivity

The great thing about both Palm and Pocket PC is they both have connectibility to a Windows PC. We have not done any test or usage other than with Windows-based platforms. Palm does its connectivity with their HotSync software, and Pocket PC has its own programs called ActiveSync. Each of these is used to connect the individual devices usually, and mostly using their respective cradles to the personal computer.

Connecting the devices to a desktop computer essentially allows them to share information and to synchronize the data between them. Setting up each is fairly simple, and once done, ongoing synchronization is quite easy. The underlying connection and the program, which passes the information between the PC and the device, is called a *conduit*. We'll discuss more on conduits later in the book.

HotSync and ActiveSync

Each of these programs is usually used via the cradles, but can be used via a communication cable. The Palm HotSync cradle is connected directly to the USB port (or serial port depending on the cable you have) on the desktop or laptop. Activating Palm's HotSync is as easy as pressing the special HotSync button on the cradle.

Simply set up the Palm desktop program, connect the cradle to the desktop computer, and press the HotSync button. Figure 1.2 shows the Palm desktop custom menu. It has all the different conduits, programs to connect specific applications, and each is customizable depending on how the user wants to synchronize. If you don't want to do anything for a specific application such as mail, choose that action and the next HotSync you perform will not copy mail from PC to Palm, or vice versa. This goes for each application on the device.

Palm also provides a special Desktop application. Most people using the Windows platform have Microsoft Outlook and prefer to connect to it. However, for those who'd prefer to use another application, Palm supplies its Palm Desktop. Date scheduling, addresses, to-do lists, memos, and more are all available through this desktop application (see Figure 1.3). You can choose to use it and synchronize to it as you see fit.

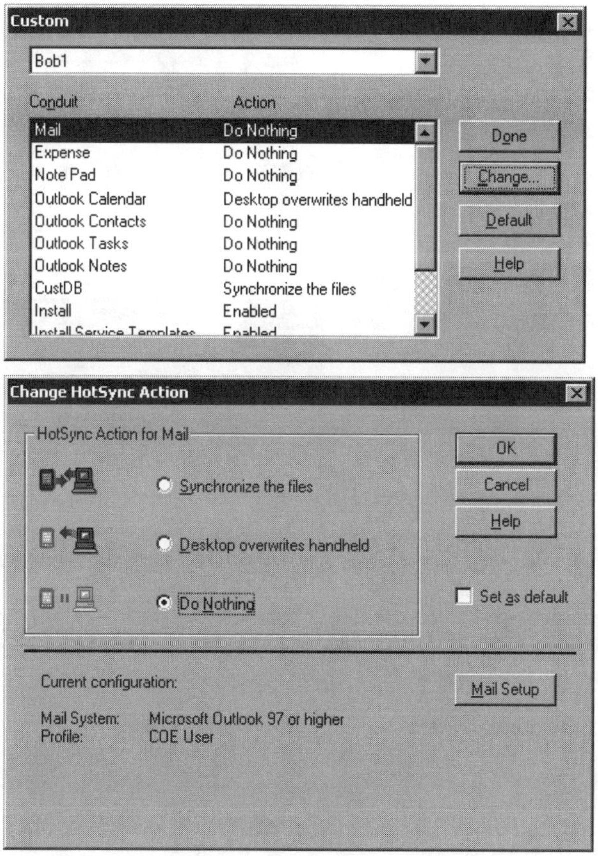

Figure 1.2 HotSync custom menu.

Microsoft has an application of the same concept, but with an automated synchronization process. For Pocket PCs, simply install Microsoft ActiveSync on the desktop computer, connect the cradle to the USB port on the machine, configure the software, and place the iPaq in the cradle. ActiveSync will automatically recognize and detect the device, and then automatically synchronize the unit with the desktop. It can also be programmed to do this on an ongoing periodic basis. Very handy!

Figure 1.4 shows Microsoft's ActiveSync application. It is the same concept as with the Palm conduit program, but represented slightly differently. This application shows the programs that ActiveSync is prepared to consider for synchronization. Again, each program has the specific option of how to synchronize, when to synchronize, and what to synchronize. There's a lot more to the program, but we'll cover the details later in the book. One nice feature to

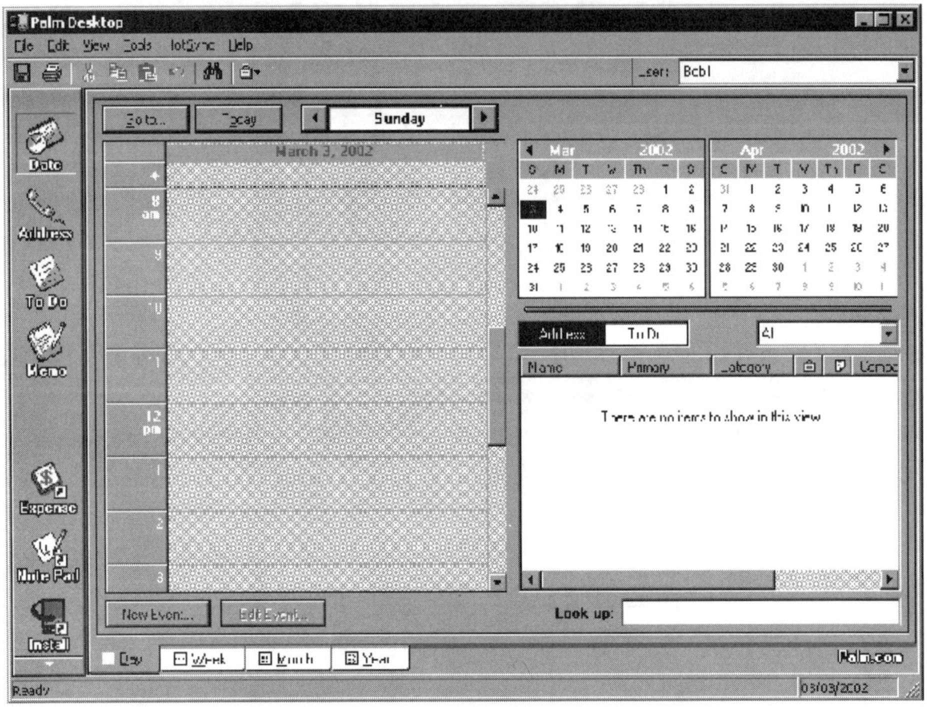

Figure 1.3 Palm Desktop menu.

note on ActiveSync is that while the device is connected to the PC and the PC is connected to the Internet, pass-through mode allows the device to access the Internet directly, making downloading very simple.

There's much more to say about these units, but let's cover that as we move forward in the book. As we explain the examples, we discuss unit features and functionality.

Modems

Part of the beauty of having a portable mobile wireless PDA is the capability to connect to the Internet at anytime. To do this, you must have some type of modem. These devices allow the individual to connect to the back-end enterprise data warehouse or operational system.

From a planning point of view, if the user connects now and then, the itemized information on the backend application must somehow synchronize with the units. This can be done by queuing the messages using some type of middleware system—obviously, some level of advanced systems architecture is required at this point. This is a much more advanced topic, but the point is that

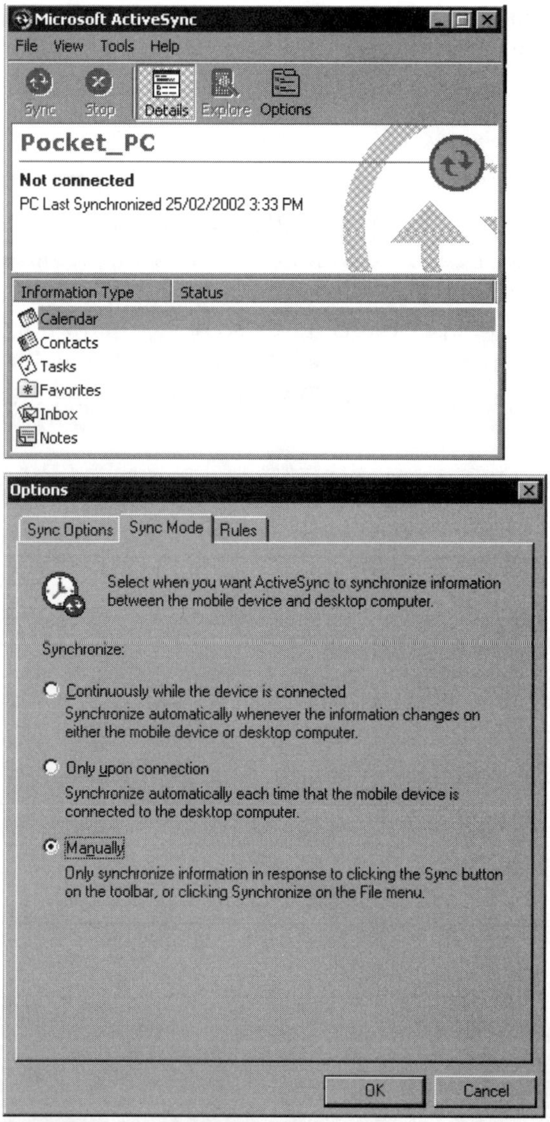

Figure 1.4 ActiveSync menu.

wireless computing with these PDA mini computers can easily connect to an enterprise system for complete data synchronization to back-end information while on the road anytime, anywhere.

Modems come in different styles. One example is the CompactFlash Fax modem. It runs at 56K and works just like a PCMCIA card but is about one-third the size, which makes it ideal for palm-size and handheld PCs. It adds little weight to the unit (one-third ounce, or eight grams), and has no impact on the size of the unit because it slides into the existing expansion slot.

Final Thoughts

While wireless Internet has come a long way, it's still in its infancy. WAP was the initial kick-start into the wireless world for computers and, in our opinion, the next step is with these PDA devices. To dive into this PDA world, the first step is learning how to build applications on these devices and how to synchronize them with databases on servers. The following chapters will walk you through the components required to build PDA applications with portable and synchronizable databases.

CHAPTER 2

Database Architecture

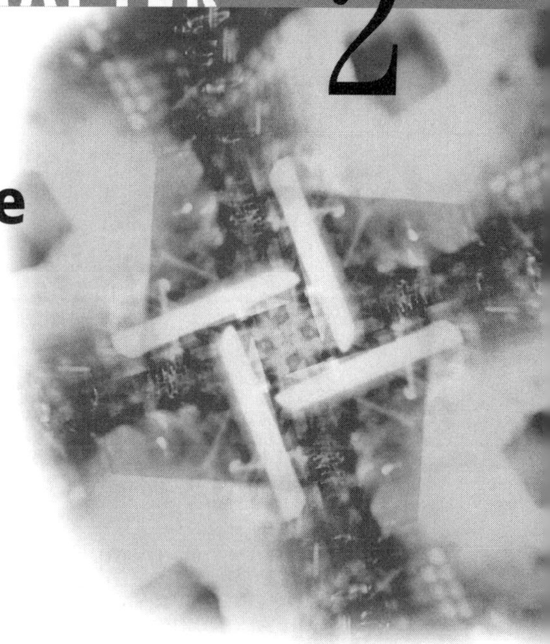

Computers have been highly used in most businesses for 20 years or more, so we feel fairly confident that most people in the workforce today have heard about databases. But let's recap to be on the safe side.

Database Basics

Databases are essentially organized, structured, and planned groupings of data files, just like files in a filing cabinet. Each file contains specific records or entries. Each record has one or more fields just like lines on a paper in a file in a filing cabinet. To access the data quickly rather than go through every file in the filing cabinet, the files have tabs with easy reference information about that particular file. In a human resource department, the filing cabinet may contain a file on every person working in the company. It would therefore make sense that the tab on these files contains each person's name. For added efficiency, the tab would have each person's last name followed by their first name since there are probably more unique last names than first names. We now have a highly organized filing cabinet with all employee files organized by last name first and first name last.

If the company was quite large, and the employee files consumed several cabinets and each cabinet had many drawers, it would be efficient to label each drawer alphabetically. So all employees with last names beginning with "A" would be in the first drawer and all employees with last names beginning with the letter "Z" would be in the last drawer. We might even organize the first drawer to contain all employees with last names beginning with A through C inclusive. What we've done here is organized our filing cabinet into sections. This can continue on into clever filing methods that may all depend on the number of employee files, who's doing the filing, who's retrieving the files and how often the files are accessed. Maybe a handful of employees have a grievance with the firm and, therefore, their files have been pulled and placed in a special drawer. This is another method of quickly retrieving specific files.

The same department may have other types of files in the same or different filing cabinets. Let's say we need to keep a file on each company-owned truck. And then in either the employee file or the truck file we need to record which employee is assigned or responsible for which truck. Perhaps this may change on a daily basis so we decide to keep an employee-truck cross-reference file. Every time an employee is assigned a truck, be it for one day or several weeks, this new file will contain the employee's name and the truck assigned. The question now is, will we tend to look up the truck information and then look up who was assigned to the truck? Or would we normally look up the employee information and which truck he or she was assigned to during a particular time period? This cross-reference may go either way depending on the circumstances and requirements of the business question being asked. This situation brings about the importance of integrity rules. We sure don't want to look up employee Ralph and find him assigned to truck number 77 last Tuesday, and then look up truck 77 for the same day to see that it was assigned to a totally different employee named Rick. As you can see, data integrity is very important and high on the requirements of all businesses large or small.

Now comes the question of security. Just who's allowed to look at these employee and truck files? Can Rick look up who was assigned to his truck before him? Probably. Once Rick sees that Ralph was previously assigned the same truck can he then look up how much Ralph traveled with that truck? Maybe. How about if Rick wants to find out how much Ralph was paid while assigned the truck? Most likely not. So, therefore, we have questions on privacy and security. Imagine what would happen if each employee knew the salaries of each of the other employees in the company. That would certainly bring about some intense situations, as each would want as much if not more pay than the others.

After all the basic concerns are dealt with, we're in business. We have our filing cabinets full of files of each employee and all trucks well secured under lock and key, only accessible by specifically authorized individuals allowed to look and analyze these files and the information therein. Great, but from a business point of view, what if the place burns to the ground one night due to bad electrical wiring? All the files are lost, including all records of who was assigned which truck, and how long each employee drove the truck. These employees want their paychecks, but you have no way of knowing who worked how long and on which truck because all the files are lost. The employees want and need their moneys due. They have families and bills to pay. They must have some type of compensation, but you're completely lost without your records. This brings about backups and a plan to reinstall those backed-up files. When you think about it, you really need some sort of plan to say when to make the backups and what to do with them once you get them, as in where you'll keep them. Then, as in the emergency scenario just explained, a plan detailing how to get them back from, say, the bank vault in a reasonable time so your business won't miss a beat. There might be some business slow down, but the business should still move forward as usual to ensure continuity and minimize customer/client loss.

You can easily see how one thing leads to another and eventually an entire system is required. As the Modern Age came about and companies became larger and more complex, people began realizing that an organized, efficient, and fast method of retrieving information was indeed very valuable to business. If we could run payroll in two hours rather than in three days, surely we'd be cutting costs. In walks the Computer Age, and problem solved. At first, just getting simple spreadsheets with quick and, more importantly, correct tabulations was heavenly. Days of number crunching performed in minutes or hours and without errors had finally arrived. However, there was a drawback: We now had to hire extra staff to create electronic filing cabinets with methods to enter and extract the data. The IT department was born.

Let's take a side step for a moment and recap the Computer Age as it pertains to databases. The filing cabinets became databases. Each file type in the file cabinet became a table. The employee files became the employee table, the truck files became the truck table, and the cross-reference files became tables too. The tabs on each file—remember how the employee file tabs had last name then first name written on each of them—well, these evolved into special indexes on the tables. Therefore, the employee table has a last name, first name index, and there might be a second index on the same table for those employees with pending grievances as in the example earlier. Not all employees would be indexed in this case, and this is the beauty of it all. In this case, if you'd want to look up employees with grievances, using this special index would result in only those people associated with pending grievances. It would be a quick, efficient, and very flexible retrieval system. You can also add or remove individuals from this type of index automatically by means of a special program the IT person created. All in all, a wonderful advancement, wouldn't you say? We could now add special rules for integrity purposes. This would resolve the issue of Ralph being assigned to truck number 77 on Tuesday, and truck number 77 reporting Ralph and only Ralph as the assignee last Tuesday, all being performed under the covers by the database itself. No more miscalculations due to improper record keeping. This does get more complicated, but the point is the simplicity, which is very feasible, and therefore, an automated database is quite the advancement in daily business functions.

The Relational Database and Its Components

There are many types of database models, such as Dimensional, Network, Hierarchical, Relational, and several others (see Figure 2.1). These are simply known as DBMS, or DataBase Management Systems. Each has its own underlying method of connecting each table and record to the next, and so each has its own naming conventions. The most popular and most efficient under general terms is the Relational database.

This type of database has certain basic terms. The grouping of specific files such as employee files is called a *table*. Each employee file itself is called a *row*. The employee table is made up of many employee rows, one per employee. Databases are essentially one big continuous file organized in sections (blocks, pages, segments, or whatever you might want to call them depending on what's being discussed). The database knows how big or long each section is, based on the size we initially allocate. The allocation is based on the size of

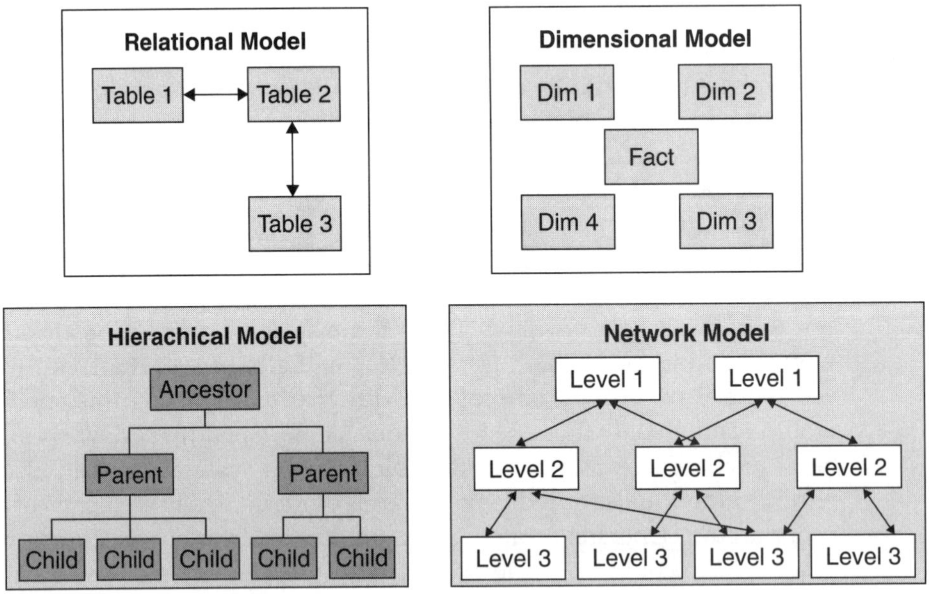

Figure 2.1 Database types.

each table, which is based on the size of each row within the table and on the number of rows in the table. Then, to further understand the components of each row, we subdivide the individual employee rows into columns. This gives us nice, easy-to-comprehend rows and columns. For example, a row for a particular employee might contain the following columns: last name, first name, and telephone number (Jones, Ralph, 555-1212).

Tables

Tables 2.1 and 2.2 show a grouping of tables within a database. Table 2.1, the Employee table, was just described in the preceding section. Table 2.2 is the Truck table, and each row represents data. There are five trucks in our fictitious company, and each has its own row in the truck table. Each row is further subdivided into columns: Plate, which shows the truck's license plate number, Type of truck, and Year, which shows the year the truck was made.

From a technical point of view, there are quite a few things going on behind the scenes. First, specific databases have specific page sizes. We mentioned before that databases are basically continuous files with data written bit after bit. The files are chopped up into specific sections called *pages* in this case. It's kind of like formatting a diskette. When a database is created, it is allocated to a specific size, let's say 10MB for our example. 10MB is the total size of the file. The file is then formatted into pages for the database. Each database has its own specific page size requirements, but most are using 2K or 4K pages these days. This is another area where it could easily get more complicated. Professional database administrators can actually change the page size to optimize the database for specific business and data requirements.

So, now we have pages in 4K (4096 bytes) sizes. 10MB (10,240,000 bytes) in 4K chunks gives us 2500 pages. Some of these pages are going to be used by the database management system for internal administration and tracking of the remaining pages as well as other tasks. Each page is then further subdivided. Each page has a header and footer. All pages are basically strung together: page 1, then page 2, followed by page 3, and so on. Each page has its own number in the header.

Table 2.1 Employee Table

LASTNAME	FIRSTNAME	TELEPHONE
Jones	Ralph	555-1212
Barns	Rick	555-8888
Jacobs	Abraham	555-1234
Ryan	Jack	555-0001

Table 2.2 Truck Table

PLATE	TYPE	YEAR
ABC123	Moving	1998
EDF456	Flat bed	2000
XYZ001	Delivery	1977
STU002	One-ton	1988
MNO003	Pickup	2000

While each database system is different, the footer basically says where each row on the page is located. There's more to all this, but let's keep it simple.

Now we tell the database that we want to create a table called *Truck*. The database gives us a specific portion of the 2500 pages. Most databases try to handle eight pages at once; this eight-page allocation is called a *segment*. If we need eight pages or not, that's what we get. We have five trucks in the company, which gives us five truck table rows. For this particular table, from Table 2.2, each row contains the Plate (6 bytes) column, Type (10 bytes), and Year (4 bytes long; for example, 1997) for a total of 6+10+4=20 bytes. For five rows, this gives 5*20=100 bytes. 100 bytes can certainly fit on one page, as each page has 4096 bytes minus whatever is required in the header and footer. Now, if eight pages are allocated automatically per table, then we have seven pages basically unused but allocated to this table. As we said earlier, some databases always allocate in eight-page segments (or chunks) for performance reasons. This is more practical when allocating database space for large amounts of data. Our example here is rather small, but you get the picture. For the employee table, another eight 4K pages is allocated and if it's filled up, a second segment is allocated. Some databases like to allocate all associated segments together, while others just allocate them on-the-fly. It's not unusual for the employee table's first segment to be allocated, then the truck table segment (eight pages), followed by a second segment of eight pages for the employee again. A bit messy, but that's what happens under the hood in some cases.

Too much information and not enough reasons—don't worry too much about it, and welcome to the world of database administration. In the old days, we had to calculate space requirements because we didn't have gigabytes and terabytes of available disk space at our fingertips. We used to forecast and calculate every last byte of available space required for the next year or two or three. Today, the database vendors are simplifying things by incorporating all the internal calculations within the database management system. Life sure got a lot easier in the past decade. Mind you, networks got a lot more confusing, and tuning is now focused on large amounts of data traveling here and there rather than in back-end databases. This is a very loaded statement, so just take it at par without dependencies for now.

We can move data faster and we can organize and hold more data, but in the small PDA world, we must again worry about database sizes, table and row estimates, required data, wanted data, and nice-to-have data.

Primary Keys

Getting back to tables and data therein, with lots of data we need to quickly identify which row is which. This is done with something called a *key*.

Continuing with our truck example, how do we identify each of the five different trucks quickly? The answer is with a specific key unique to each row. We could have numbers on each truck, like 77, 55, 33, 11, and 99. We're going to leave out modeling issues here and talk about unique identifiers with no business dependencies. All we need to know at this point is that each row has its own identifier called a key. To be more specific, these are called *primary keys* to the table.

If you want to access information on truck number 77 such as each individual column, you'd simply say: Select ID, Plate, Type, Year from Truck where ID=77. It stands to reason then that the primary key must be unique. What good would a key be if it retrieved multiple rows? The purpose would be lost. Therefore, a fundamental rule here is that primary keys must be unique and identify one and only one row, no matter how many rows exist or can possibly exist in a particular table. Another table can have its own primary key, which might be the same column name and value, but it would be unique to that table.

With the addition of the identifier, our table now has an added column called ID (or whatever) as shown in Tables 2.3 and 2.4, which changes our page size estimates. For the small number of rows in this particular table, we're not too worried. However, in large databases with gigabytes or terabytes of information, every new column is extremely important, and the number of rows overall for that table must be taken into account. Databases for small devices are

Table 2.3 Employee Table

ID	LASTNAME	FIRSTNAME	TELEPHONE
01	Jones	Ralph	555-1212
02	Barns	Rick	555-8888
03	Jacobs	Abraham	555-1234
04	Ryan	Jack	555-0001

Table 2.4 Truck Table with Primary Key

ID	PLATE	TYPE	YEAR
77	ABC123	Moving	1998
55	EDF456	Flat Bed	2000
33	XYZ001	Delivery	1977
11	STU002	One Ton	1988
99	MNO003	Pickup	2000

very similar to large databases in that space restrictions and consideration are of high importance. In our case, we have to take into consideration every single bit of data because PDA devices have limited space. Sure, we can add flash cards and obtain five gigabytes of available space, but not always. If we're limited to 32MB of data and we want to add lots of data to the device, we must really be aware of all data, keys, indexes, and everything that takes up disk space.

Indexes

On to indexes! We have a database with tables consisting of rows with columns. Primary keys and foreign keys are columns that must have indexes associated with them. Indexes can also be defined individually on one or more specific columns, usually because the database designer believes that most queries will be using these specific columns to obtain information from the table.

A *primary key* is the definition of one or more specific columns on a table that will be used to uniquely identify rows. It's the index behind the scenes that really allows for the fast accessing of the row based on the predefined columns. In databases, when a query is executed, the database engine will try to figure out if a predefined index matches the query. This gets complicated, so we'll just assume that the database knows what it's doing. The point is, other than really understanding how the database figures out which index to use, you really don't have much say in how the data is being accessed. However, when you specify the primary key in the query, you're specifically telling the database to use the underlying primary key index. This really overly simplifies the process, but you get the picture, right? Same idea behind foreign keys, which we'll discuss in the next section.

Indexes are arranged in a structured fashion allowing for faster lookups. One could simply flip through each page in a book until he found the sought-after page, but that would be very time consuming. Database indexes are just like indexes at the back of books. Look for what you want in the book index and you'll find exactly what page it's on. Flip to that page, and there you go.

Indexes are made in different ways to ensure the optimum method of getting to the required data. The different types include Bitmap and Btrees. At this point, you just have to know how the data is going to be accessed, and build indexes that best suit your requirements. In our employee example, we want to look up people based on their last name and then their first name. We simply create an index on last name followed by first name. Done. If we only wanted to use the person's last name in a query such as select lastname, firstname from employee where lastname='Jones', the lastname and firstname index would be used because lastname is the first entry in the index. If we run another query using both lastname and firstname, the same index would be used because the first index entry is used. However, if we only specify the firstname, chances are that the index would not be used because the lastname is the leading (first) column in the index and therefore must be specified if you want that index to be used. This means that a query specifying only the firstname would be much slower than the same query using the lastname in the where clause—depending on the amount of rows in the table, of course. Consider looking up a person in the telephone book. If you only know the person's first name, you're out of luck, because the book is organized by last name first and first name last.

In conclusion, indexes are quite important to databases. Primary keys are, essentially, user-assessable indexes and, therefore, usually ensure fast query response if used.

Referential Integrity (Foreign Keys)

Remember the data integrity issue mentioned earlier? Well, this is where we take care of that. If we have a primary key defined on a table, we are ensured that we have one unique identifier on that table. If an employee row were added to the Employee table, we know it would be unique or it wouldn't get added. Same idea for the Truck table, add a primary key and we ensure that there is one unique truck identifier in the entire Truck table no matter how many rows are in the table. Now, if we want to cross reference employees with trucks, we get the integrity issues mentioned earlier. We need a method of ensuring that if Ralph was assigned to truck number 77 last Tuesday, that truck 77's point of view shows that Ralph was indeed the assignee and not another employee.

To resolve this issue, we create a new third table called Employee-Truck-CrossReference table. This table will keep track of all trucks assigned to employee drivers. For business and practical purposes, we could rename the table to DriversAssigned, as per Table 2.5.

The employee ID in this new table relates directly back to the ID in the Employee table (see Table 2.3). The column names don't have to match, as the important

Table 2.5 DriversAssigned Table

EMPLOYEEID	TRUCKID	DATEASSIGNED
01	77	2002-05-02
01	55	2002-05-03
02	77	2002-05-01

point is that the relationship exists. This relationship is called a *foreign key* relationship. The EmployeeID column is a foreign key to the primary key on the Employee table.

Think of this relationship as a parent-child relationship. The parent must exist for the child to exist. There can be no child without a designated parent. The EmployeeID column from the DriversAssigned cannot have a value unless that value is already in the Employee table in the ID column. This process is called *data integrity enforcement*. The same thing happens with the TruckID in DriversAssigned and the ID column in the Truck table. Now we're sure that data integrity is enforced at the column level. Data integrity is usually referred to as *referential integrity*. Since foreign keys refer to primary keys, the relationship has been defined as *referential integrity*.

However, we still haven't resolved the matter of enforcing one driver per truck at any particular time as per our earlier issue. The matter now comes down to enforcing one driver per truck per day. To do this, we add a primary key to the DriversAssigned table, which includes all three columns. Since primary keys are unique per table, we can then only have one driver for one truck on one day. Now all the rules are enforced and we're sure that only Ralph was assigned truck number 77 last Tuesday, and from truck 77's point of view, it was only assigned to Ralph on the day in question.

Figure 2.2 shows the relationships between all three tables. The Employee table has its own primary key and index on column ID. The Truck table has its own primary key and index on its own column called ID. Same column names, but that's okay because the names are unique on the individual tables themselves. The third table, DriversAssigned, has its own primary key made up of all three columns called EmployeeID, TruckID, and DateAssigned. We could have called the ID columns something else, but for the sake of simplicity, we called them the same as their originating table names plus the ID word. We then define these same columns as foreign keys to the other tables. This ensures that we cannot add a row to the DriversAssigned table if either of the primary key values is not initially in the originating tables, called *parent* tables. We couldn't add a row to DriversAssigned with an EmployeeID that doesn't exist in the Employee table, and we couldn't add a row to the DriversAssigned

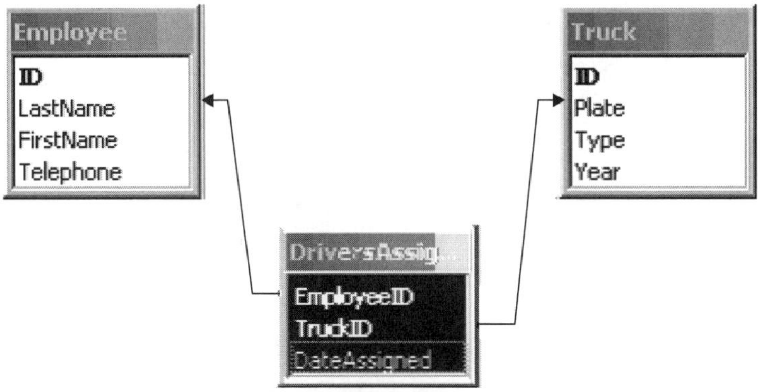

Figure 2.2 Referential integrity on tables.

table if the TruckID were not already in the Truck table. This then enforces data integrity between the DriversAssigned, Employee, and Truck tables.

Joins

This is an excellent time to discuss *joins*. The referential integrity, foreign keys, in our ongoing example, describes a middle table pointing to two other tables as it represents the relationship between each. From a practical business usage, it's very reasonable to request information such as a list of all the drivers who have been assigned to truck 77. The query would be written as follows with the following join criteria:

```
Select lastname, firstname
from Employee, DriversAssigned
where
Employee.ID=DriversAssigned.EmployeeID and
DriversAssigned.TruckID=77
```

Both Employee and DriversAssigned tables were specified. The join criteria in the query was done on the Employee's ID column and the DriversAssigned EmployeeID column. We didn't have to look at the Truck table because we knew the ID and therefore simply used it as a qualifier in the query. We could enhance the query and request the year of the truck by doing a second join between the DriversAssigned table and the Truck table using the TruckID column from DriversAssigned and the ID column from the Truck table. This type of join is called an *inner join* or just *join*. Think of two tables intersecting. We want certain column values common to each of the joining tables.

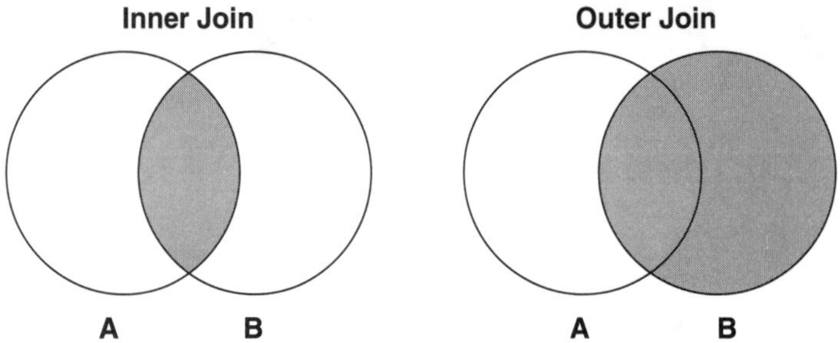

Figure 2.3 Join types.

Outer joins are another type of join. Outer joins are queries, which want the common column values between two or more tables and still want other rows from one table not found from the join criteria. Results from the join are then all rows in the intersection plus other columns from the outer joined table.

From the diagram in Figure 2.3, for an outer join, you might ask, "why not simply select all columns in table B?" The answer is because you might want one or more columns from A, but these columns might not be available for rows in B that are not in A.

Cursors

Armed with the preceding information, we can now design, load, and retrieve all the information in our database. However, what if we want to perform some sort of process for each row returned from our query? Let's say we get two rows back from our previous query, which gave a list of all drivers who have been assigned to truck 77. We'd need to somehow stop the processing after each row and execute another procedure. This is done with the help of a *cursor*.

Basically, we're going into a sort of holding state per returned row from our database query. As we loop through the result set, two rows in this case (one for Ralph and another for Rick), we can update another table. This is an overly simplified example and it could be done in one snappy query, but for the sake of our example, we'll stick to updating table X, which we won't even bother to define at this point.

In general, a cursor is defined for a particular query and given a name. For example:

Define Cursor ABC for
Select lastname, firstname
from Employee, DriversAssigned
where
Employee.ID=DriversAssigned.EmployeeID and
DriversAssigned.TruckID=77

Then, later in the program, we'd run the cursor and use it as a loop. For example:

For each ABC
execute UpdateProcedure

When finished, we close the cursor, because cursors hold ongoing internal counters and pointers within the database management system. The close might look something like a simple

Close ABC

That's it, now you can perform subprocessing for each returned row from your database query as the program loops through each returned database row.

SQL

SQL is short for Structured Query Language. This language contains the Select statement you read about earlier. It also encompasses other commands you'll become aware of as we move forward. Each database has its own set of SQL commands. A special committee has standardized SQL worldwide, but there are differences from vendor to vendor. The standardization means that as different versions of SQL emerge (since it is versioned just like any software package), the language has basic functionality no matter which vendor produces the SQL. An example of basic functionality would be the way results are sorted (order clause in the Select command, for example). How joins are done—or should we say "written"—in the language is another example. The underlying database might perform the join in a special way that could be much more efficient than another database vendor's software, and therefore a huge selling feature.

The following is a history of SQL standards, which database vendors currently use. We won't go into the differences between these, as there are many.

- 1986 (SQL1 or SQL-86)
- 1992 (SQL2 or SQL-92)
- 2000 (SQL3 or SQL-99)

Databases on PDAs are small because these devices are still limited in power and disk space. The database management systems on them are very restricted; therefore, the database access language (SQL) is also restricted. Most of these databases, as you'll see later in the book, have very small footprints, meaning that they take up very little space on the devices. To do this, the vendors stripped down the database functionality and limited the SQL features and commands to the bare essentials. As the individual databases for PDAs evolve, the SQL functionality will grow to one day be equal to those on personal computers and professional versions.

Security

As mentioned earlier, security is a real issue. We certainly don't want Ralph to view Rick's salary, and vice versa. Imagine some stranger being able to see your bank account information anytime he or she wished. Worse, imagine if anyone could view your personal information such as address and telephone number. We certainly wouldn't want that.

Each database must, as a fundamental function, allow security and privacy functionality. Individual users must be definable, and restrictions on access to individual tables must be enforceable. In current databases (non-PDA versions), users can generally be grouped into business units, administrator categories, and so on. This allows for easy enforcement, as everyone in the same group can perform the same functions since they would all have the same level of authority. If need be, we could define rules to the individual users, but these should be exceptions to the rule. This holds true for portable mobile systems. We must be able to restrict usage to just the individual users on PDAs, but also most importantly, restrict contact with the legacy database on the host computer. If a user tries to download confidential information to his or her PDA, that information is now available anywhere at anytime. The person could then easily transmit the information to unscrupulous competitors. If the unit were lost, anyone could easily have access to its information. Security is very much an issue in data accessibility and in the ongoing usage of the information.

All PDA databases have the capability to synchronize with the host computer. Synchronization might involve the PDA information overriding the host information, or vice versa. Imagine the difficulties if a salesperson unknowingly uploaded corrupted data. The ramifications could be enormous. Viruses are another issue. Just because you're using a PDA doesn't mean you are not open to viruses. We'll be seeing a lot more of these as portable mobile usage becomes more popular in the years to come.

Database Administration

In a typical corporate IT project, the end users, or client, request (and pay for) a database and application. The business analyst will gather business requirements from these users and relay the information to the data modeler. Data modelers will design logical models of what a database should look like based on the business and data requirements. Once completed, the database person will take the logical model, optimize it for actual usage and physically create it. This involves adding indexes based on usage requirements relayed to him/her by the business analyst and the data modeler. Once completed, the database environment is ready for the developers to begin writing programs to manipulate and retrieve data into the physical model. The end product is then returned to the end user who initially began the process in the first place.

The database person plays a very important role, as this is the person who actually builds the physical environment. Many times, for smaller projects, the database person is the business analyst, data modeler, database modeler, and programmer all rolled up in one. The fewer cooks in the kitchen, the better, right? Well, that depends. The cook had better have lots of experience in every area, on many different dishes, and under many different scenarios, or you end up with a very odd-looking cake. We've seen too many projects use a great programmer to define the models. The person might be a great programmer, but he might have no concept of the importance of proper modeling. Two months down the road, the database ends up being good for only one program and cannot be easily enhanced. A good database design allows for flexibility in usage and is the key to any project. Build with expansion and flexibility in mind and you'll always have a great system.

At the same time, the database person must also know how to operate and administer the database. He must know the internals of the database, the data usage, and the business requirements. This will then allow the addition of proper indexes, perhaps placement of certain tables close to others while separated from some completely different ones for performance reasons. Again, synchronization and security issues arise.

The administration of a database encompasses installation, upkeep, and ongoing maintenance. For a mobile environment, this also involves ensuring the proper synchronization of data between the PDA and the main database. This can be a very complicated point if left to chance. Therefore, plan, plan, and plan a strategy of moving data up and down between the platforms. Decide what to upload and download, the frequency, and when to initiate the process. These are highly dependent on the application and on system availability, which is dependent on connectivity.

As part of the strategy, there are basic issues such as how to start the databases on both ends. This sounds a bit simplistic, but the thing is, if you can't figure out how to start the program with all the parameters and dependencies, you might not have a solid environment. Sure, all might be great for a while, but administration is very concerned about failure. We do many things just to ensure that if a failure does happen, we can recover and get all information recovered so business can continue in a reasonably fast timeframe. Data is the major asset of every business. Lose your computer and your business might fail. Imagine losing your wallet or purse—that's a scary thought. Now imagine losing all your company data—you might end up without a job.

The next step is to connect to the database. How does one actually connect to the database? If you've ever been unable to connect to your data, you know what we're saying here. It's a pain. You don't know why, you can't figure out what's happening since you've been following the same simple steps for who knows how long, and yet all of a sudden, you can't access the data. Whom do you call, how long will this take—you're stuck. End users are usually non-technical, and when they can't continue with business due to a technical error, you have problems. In the mobile PDA world, the database administrator is probably the main contact for all problems with the PDA application. In this case, the person must be aware of the application and database. The person must know how to start the database, how to connect, how to access the database, and finally, how to disconnect the application from the database.

Backups

An important portion of database administration is to ensure data availability and safekeeping. The operational database on the PC or network has its own backup and restore strategy, hopefully, and now we must take into consideration the data on the PDA.

Under normal circumstances, the user should be able to connect to the network or PC via HotSync or ActiveSync (whatever conduit tool is being used). The data from the PDA can then be synchronized and backed up on the PC. However, what if the connection cannot be made? If the wired or wireless connection is not available, there must be a backup plan to save the data. We could just hope for eventual connectivity, or we could be a bit more practical and plan a strategy, which could include storing the information on a memory card (see Figure 2.4). Most new devices have the capability to insert additional memory via a memory card or expansion pack. These are the best places to store the database and application, as the card could be removed and saved in a safe area.

As a note, be sure that old databases of information match up with the application. Many times, the data is backed up and then the application changes. If

Figure 2.4 Memory cards.

you have to restore an old version of the database, the application might not be able to access it because it is no longer compatible. This happens quite often after upgrading software. This is not only a PDA issue, as it happens with large databases as well.

PDA Database Considerations

Since PDA devices have relatively small memories, the design of the applications and databases must be optimized. Older devices only have 8MB of memory, but the newer versions usually have a minimum of 32MB. Newer models have the added feature of a PCMCIA card, which can add up to 5GB of added memory. This really opens up the application possibilities and, therefore, the usage potential.

Since these devices have small memories, the database vendors had to figure out a way to implement a database on the units, which allowed for efficiency in query processing and yet didn't consume all the available memory. As you'll see later in this book, each vendor took a particular approach. Most limited the SQL features, which reduced database size, while others decided that pre-specifying the exact queries and then building database tables with only columns required by the queries would reduce the database footprint. Still others decided to implement each table separately as its own file.

Final Thoughts

From a database design point of view for PDAs, we highly suggest to only plan on using exactly what is required. Do not build extra columns if they will not be used. Plan every column usage and know your data. Determine how the data will be imported and/or exported to the driving database on the server as part of the database strategy. Plan your synchronization strategy as part of the database design phase and your PDA database application will hopefully be a success.

To understand more on the two components of the PDA client and a database on the server, the next chapter will explain the issues and considerations for PDA applications.

CHAPTER 3

Client-Server Architecture

This chapter explains and pinpoints the architectural components of using a PDA device as an extension to an existing (or new) server-based database application system. With the enterprise moving into more access points to corporate data, an understanding of client-server architecture is a fundamental stepping-stone in the development process.

Client-Server History

A major step in the advancement of computer technology was the database. The transition from flat files to a structured, organized, and optimized method of storing and retrieving information was a tremendous move forward. This was way before the personal computer came about, before the Internet, and long before the wireless computer was even conceived. It was the first attempt at understanding data architecture and modeling.

In the beginning of the Computer Age, computers were primarily used for processing and calculation purposes. They received input data, processed the data, and returned the results. With the advent of databases came the retention of input, processes, and results. Databases were storing scientific data, accounting entries, and the like. Repositories of information were in their infancy and growing. Soon, processing systems were highly dependent on pre-stored information in these quasi-architected databases. We say quasi-architected because the science of data modeling was in its infancy, since it had just been discovered.

Remember the database types model from Chapter 2, "Database Architecture" (Figure 2.1)? Well, the first databases were usually modeled in the hierarchical fashion. This meant chaining data records together, but under owner tags called *parents*. This was still the flat file layout, but with individual blocks chained together, making the entire architecture much more organized and modular. This allowed for faster processing than flat files did. With flat files, the program would begin to read at the start of the file and proceed until it found the data or reached the end of the file. With hierarchical databases, the program could determine the required ancestor, then loop through the parents, and then onward to the child records for each parent. Once the proper child record was found, the job was done. This is a simplistic but advanced view of databases in those days. It allowed for much faster processing and was hailed as a marvel of backroom computer technology.

Now, all this seems fine and dandy, but the real kicker was the fact that multiple users could now access the same data at the same time. Think about this. Only about 20 to 25 years ago, large corporations were running their entire computer systems and divisions, mostly for accounting departments, on flat files. For those of you who remember those days, computer programs were all based on punch cards. Flat files were simply one record per punch card. Very large computers read the cards until the card being sought was found, and then retained the ongoing calculation results in memory if memory was available—which is laughable from today's standards since we have huge memory space available and the capability to create temporary or permanent files directly to disk. Each program had its own set of punch cards, and all input was also retained in punch cards—can you imagine that!!

When databases came about, information was electronically retained—and with this retention came the concept of many users accessing the same data at the same time. Well, it wasn't really at the same time, since information was tagged as in-use, but only for a percentage of the time rather than the entire time a program was using the file (data).

It was at this time that things started to gain momentum. Corporate divisions were now obtaining calculations and processes computed at a much faster rate than before. The individual departments saw a reduction in staffing while the computer division saw a huge boost in new resources. The business was moving forward, making critical decisions in days rather than months. Those who had this amazing computer technology were making waves. Business users were able to ask questions of the data and input information as it happened. Data and information were quickly becoming the fastest growing technology, industry, and business drivers than ever before in the history of humankind.

There was just one drawback: All the computing had to be done from one location and fairly close to the main computer. These computers were massive.

This is fairly tough to understand unless you had worked on these early computers. Twenty years ago, a typical computer system would take up about the same amount of floor space as four to five typical living rooms. The computer rooms had central air-conditioning to keep these high-powered machines from overheating. The printers themselves had their own mini air-conditioned rooms too. The amount of disk space, memory, and processing power on today's laptop far outperforms the supercomputers of the early 1980s (don't even ask how little disk space or processing power those early computers had!!). However, this is only one side of the story. Think of the potential of today's computers. We can now run video directly from the Internet; trying to simply run a large Cobol program on a 1981 supercomputer was an amazing task, and now real streaming video is possible—wow!

But we digress. Let's get back on track. Computer science in the early 1980s was being driven by database technology—in our opinion. From this came the need to access data from more and more distant locations. Therefore, rather than having your office on the same floor as the computer room so you could run over and execute your punch card deck, the demand became more focused on access to computer programs. Actually, this was more of a formatted approach to data retrieval, from remote offices. These offices would be in buildings next door or down the street. In came network technology and with it, came the ability to divide the machines into clients asking for service from a centralized server.

What Is Client-Server?

Network technology allows for the transportation and communication of data. Think of networks as the highways, and data as the automobiles. With the fast growing requirements to access data on an ongoing basis, information highways (networks) were being built and enhanced. The big problem was moving data to the end-user's workstation, which back then was usually an IBM 3270 unit, nicknamed a *dumb terminal*. All this terminal could really do was send information messages to the computer system, usually an IBM 360 or 370 (from our experience). The computer would then decide how to process the information, whose information to process first, which user to service, and so on. Dumb terminals were based on time sharing and officially labeled as such by the acronym *TSO* or *Time Sharing Option*. The point is that all of the processing was being done and controlled by the main centralized computer system. This meant that network traffic was increasing and the computer highways were getting somewhat jammed.

Right about this time of increasing worldwide heavy usage of mainframe systems, came another revolution in computer technology called *client-server* technology. This was the concept of making the end-user's workstation intelligent—the dumb terminal finally gets a brain.

The idea here was that if more processing could be done to the data on the end-user's computer, there would be less overhead and burden on the network and on the central processor (the main computer). Okay, so what does this mean? Well first of all, the requests and processing would be broken up into pieces so different machines would handle different tasks. If the processing could mostly be done on the end-user's machine, that would theoretically leave just data with data retrieval (along with insert and delete) requests as the only network traffic to the main computer—in a general data sense. The main computer's database would then process the requests and simply pass the results to the workstation via the network. The bonus here is that the user would be using two computers instead of one. The end-user workstation would retain information since it would have increased memory. The front-end programs could run on the workstation, leaving the main computer free for the already increasing workload from the ever-increasing number of end users.

This is when we should have all invested in Microsoft, Intel, Cisco, and IBM. These people really changed the world. IBM was first trying to enhance the dumb terminal, and at the same time was building personal computers for individual usage. However, the first thought was to build a smart workstation to take the workload from the main computer. This would then benefit the end user or client, as processing power would be totally dedicated to the individual's machine. The main computer would be used to serve the client requests; hence the term *client-server*.

Around this time, give or take several years, a soon to be famous unknown and his pals were hard at work building software for these new-fangled IBM workstation computers. This software was meant to simplify usage of these still non-mainstream personal computers. His vision was to reduce the complexity of a typical computer so that everyone would, one day, be able to own and use his or her own computer. This man was Bill Gates, and his fast-rising company was called MicroSoft. And MicroSoft was growing! (A little trivia: Back then, they spelled the company name with a capital "S.")

His (Mr. Gates) idea was called Windows. Back in those days, Windows was simply a program that ran on the IBM PC. It gave the users a nice graphical representation of the data. Most notably, it consisted of file manager, a real nice and easy program allowing the user to view all files on the computer. Then came the inclusion of Lotus 1-2-3, a spreadsheet program similar to Excel, and Word. These two programs allowed the individual to use the computer as a tool for calculating and retaining ledgers or spreadsheets and writing. These little programs revolutionized the world and brought us to where we are today. Of course, a lot more was happening at the same time, but these little programs, along with the concept of opening a window to view your computer data and to multitask, were perfectly timed—at the time.

How does this all relate to client-servers? Well, while IBM was trying to get more processing power from their mainframes through re-architecting usage, Microsoft was trying to introduce this new Windows computing concept. IBM was at the center and put two and two together, coming up with the answer to their architecture issue—the personal computer in the workplace with processing power to boot. This expanded the corporate mainframe and allowed end-user departments to perform their own independent functions such as spreadsheets pertaining to their specific roles and duties. This empowered the individuals and was a lot more palatable to the departmental budgets, too. As this grew, individual home usage grew too, mostly powered by games at first, but the point is, it was a start.

Now we had client-servers, and the personal computer was hitting the mainstream. Small and medium-sized businesses could afford these fancy machines, individuals were becoming creative, and the world was changing. Networks were growing, a major movement in processing power was underway, and memory and disk capacity was hot stuff. Intel was leading the way on chips. Every year and a half, the processing power of a chip was doubling. Individuals and corporations were buying personal computers, money was flowing, research and development was growing, and chip capacity was advancing at alarming rates. Microsoft was coming out with upgrades and enhancements every six months. IBM couldn't make these machines fast enough. Students were making and selling these new personal computers from their dormitory rooms at universities—this is how DELL computers got its start—and network and database technologies were taking giant steps.

Client-servers were big! Let's look at this client-server concept a bit closer. The idea is to remove some of the processing from the central computer and replace it with the capability to execute the process on a smaller yet powerful machine. The central computer would then become a server to the client's requests. We can see this in Figure 3.1.

Mind you, a client and server could reside on the same machine. This depends on the logical organization of the user front-end application and the back-end server programs.

In effect, client-server started with the concept of reducing mainframe usage and moving the processing power to the client workstation or personal computer. It evolved from full mainframe usage to a heavy emphasis on database servers. The client program would send requests to the database management system (DBMS) residing on the server. The server would then process the request and return results to the initial client program for further processing and display.

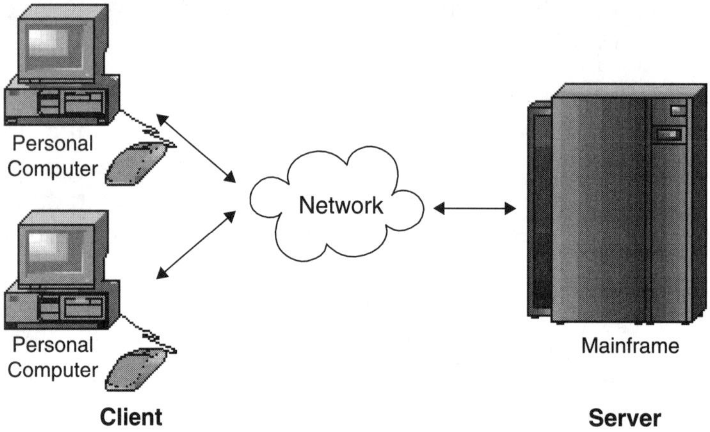

Figure 3.1 Client-server.

Fat versus Thin

The question now is, how much data and/or programs remain on the server, and how much remains on the client?

A *fat client* refers to placing much of the programs within an application on the personal workstation or computer. A *thin client* refers to having only several programs on the client with much still remaining on the server. The more programs on a machine meant more processing on that machine.

Each of these client options has its own pros and cons, each depending on the application programs, the location of the servers, and network constraints. A fat client will take up lots of disk space on the client. A thin client will gobble up less disk space on the client, but place more of a burden on the server. Propagation of upgrades and patches favor the thin client, since fewer programs would be distributed to thin clients. However, there have been advancements and new strategies in allowing the distribution of software in the past several years. From a network point of view, the thin client model is most highly preferred. Make an enhancement in one centralized area, and the server and all users obtain the same unified result of the change.

The Internet age also brought focus to thin and fat client architectures. The individual's personal computer is the client, the Internet or intranet is the network, per se, and the server is some computer that could be anywhere in the world. Most Internet-based applications are accessed via a browser such as Microsoft's

Internet Explorer or Netscape. When a page or site is accessed, the images, certain programs, and data are all sent to the client machine. However, most of the processing happens on the server side. If the program accesses a database, the database itself processes the information. Sure, JavaScript can do much processing on the PC browser side, but Visual Basic still runs on the server.

In the end, a tailored fat or thin client-server architecture will allow more improvements in usability, flexibility, interoperability, and scalability when compared to the centralized, mainframe, and time-sharing computing.

Tiers

With client-server architecture, there are many potential methods in designing programs and data. Actually, when you think about it, these components can be broken down into processing areas such as presentation layer, logic layer, and the data layer.

The presentation layer deals with how the data is viewed, assuming we're not processing data in a batch mode. Batch mode is a lengthy automated approach to processing data. Suppose you want to add up a customer's bank balance for a certain day. Take the opening daily balance, apply all the daily transactions such as deposits, subtract withdrawals such as bill payments, and add interest. The end result is a closing daily balance. Great if this is a one-time affair, but what if you were required to tally all end-of-day balances for all clients for each day in the month. Obviously, doing this by hand will be very time consuming and probably unrealistic. So instead, put all your commands in one program along with some sort of tally logic to calculate the end result and have it initiated at some off-hour like 2:00 A.M. (If you have 10 million customers, this might take a while, so these types of batch programs are usually run when no one else is using the system, which is usually sometime in the wee hours of the morning.)

A typical presentation layer these days is a Web page presented with the Microsoft Internet Explorer or Netscape browser. Sure, there are other methods and programs to present results and to gain access to back-end data, but let's think Web browsers for our presentation purposes.

The next layer is the logic. This refers to the guts of the application, which for simplicity will be a typical computer program. There could be one, two, or quite a number of programs all bunched up in a prearranged manner so as to execute pieces of work in a prearranged method. These programs are the essence of the data request and data manipulation processes. The majority of logic for the overall presentation to data extraction (or whatever) steps is within this layer.

The final step is the data layer. Can't really do much without data. This layer could be a simple flat file or an entire database management system such as DB2, SQLServer, Access, Sybase, Oracle, Excel, or whatever. The point is that the data layer is sectioned into its own instance that can be maintained independently of the other layers and, hence, managed and administered to optimize its accessibility and availability. Since one of the corporation's largest assets is its data, the databases are very valuable, and, as such, must be organized, maintained, and secured independently.

Data can be sectioned in many different ways depending on the end-users' access requirements, the database management systems configuration, frequency of use, number of users, and perhaps locations of data access. As a brief insight, Figure 3.2 shows a few ways in which data can be organized and distributed in general.

As seen in Figure 3.2, data can be replicated on multiple databases, which could be on a number of different servers. Alternatively, data could span data files, as seen in the partitioned method in Figure 3.2. Another method is the stripping of data, shown in the reorganized method. Yet another possibility could be when data is specifically cached in memory. In that scenario, particular data is highly requested, and it would be efficient from the usage point of view to have it always readily available; hence retained in cache.

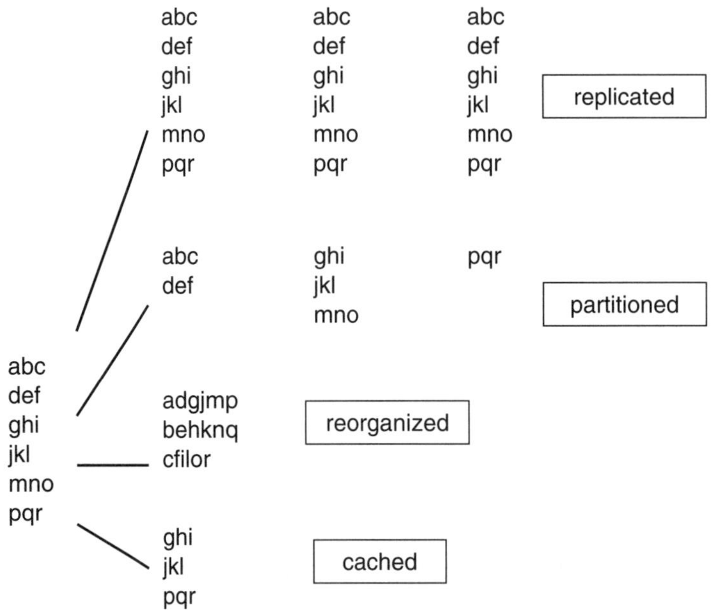

Figure 3.2 Data distribution.

At this point, we have three components: the presentation, logic, and data layers. Each layer could be divided into different combinations to satisfy different types of usage based on client-server configuration, network capacity, and device capabilities among other potentials. Figure 3.3 shows different types of setups.

In Figure 3.3, the presentation, logic, and data layers are distributed in different fashions depending on the type of network and usage. The following gives a very brief description of each type of configuration. Based on the components described earlier, Figure 3.3 shows how each section can be separated and gives a scenario to which it would pertain:

Distributed data. This may result from decentralized database servers, perhaps located in several different physical locations around the world.

Remote data. Might be due to a database server completely dislocated from the front-end presentation and logic systems.

Distributed transaction. This involves the breakup of program logic with some processing being done on one platform and some being done on the database server.

Remote presentation. The configuration is similar to a browser accessing the Internet, where logic and data reside on a Web server and the Web browser is on a client machine.

Distributed presentation. This would be similar to a partial front-end residing on your laptop but being driven by a front-end application on another machine being distributed across the network, with that machine having programs and data under its umbrella.

There are more potential configurations, and we're sure that even more will follow in the future.

The tiers of the client-server configuration and data distribution are usually threefold; hence the 3-Tier representation, as shown in Figure 3.4.

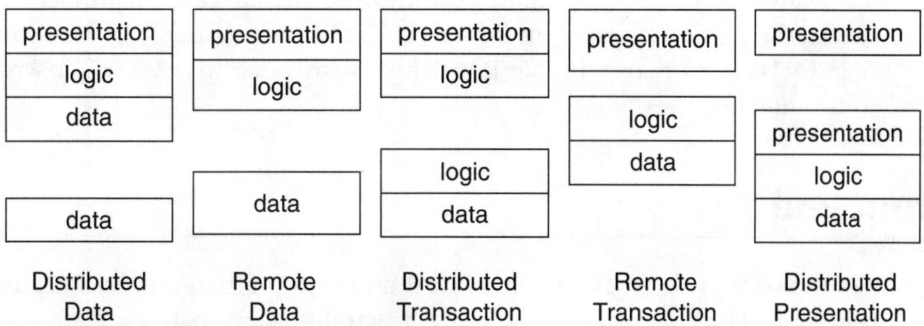

Figure 3.3 Client-server configuration.

Figure 3.4 3-Tier model.

In the 3-Tier model, the presentation layer resides on the client machine, which is usually the personal computer. The middle server would normally contain the logic process such as the Visual Basic programs. The final server would contain, manage, and organize the data in the database management system. The entire server in this scenario would, or should we say, *might* be dedicated explicitly to the data.

Many Internet Web sites are only based on a 2-Tier model. The client side is the individual's personal computer, and the server would contain ASP on an IIS server also containing an Access database. This is certainly a very popular method. Corporations, however, usually maintain, at minimum, a 3-Tier model to ensure that their assets are distinct and usually maintained by the data management department, while the applications department usually maintains the programs. The 3-Tier approach allows for simpler portability, as servers can change, data can move, and access can be redirected away from unique points of failure. Back-end systems can move to different machines, platforms, databases, and operating systems without the front-end presentation layer being affected.

No matter the number of tiers (servers), the user can read data from a site or single point of entry and not have to worry about access or location of data. It's the same idea as browsing the Internet—you really have no idea where the data comes from, where it's located, or whether it's on a single machine or distributed across multiple database back-end servers. When updating the data, the user really doesn't know where that data is located, or whether the data is replicated or not. How would you know if the credit card information you entered when purchasing an item over the Internet is really only stored in the vendor's location, or whether it was just distributed to a web of unscrupulous Internet sites?

Front End

As the back end can change in almost any fashion imaginable, so can the presentation layer. Consider a large data warehouse repository. Data warehousing is the accumulation of data upon data over time, usually in a single repository or warehouse. The data layer in this case is completely focused on gathering, capturing, and holding the same data layout (with changing values) over and over again. This could be something as simple as capturing your savings account balance every day or every week. You might access that data from a Web site application or from an ATM machine, or the bank teller might access it through special banking programs. The data remains the same, the access logic remains the same, but the front end might change. Excellent flexibility.

With wireless access, Web access, or any other method of accessing programs and data, systems are becoming very difficult to write. A programmer must ensure that all possible methods of accessing the data function correctly, and this is no easy task. We have a Web site called Worldjobmart.com, which is an Internet job site where corporations from around the world can post their job vacancies via the Internet. This was created with simple HTML in the presentation (browser) layer. When the Wireless Application Protocol (WAP) came about allowing access to Web sites from special cellular phones, we had to write a bunch of new programs just to accommodate this new HDML (Handheld Device Markup Language) language. Same database, mostly the same subroutines but different presentation programs. Then we decided to make the site available to the iMode protocol for use in Japan. Once again, this was a new presentation layer.

The HDML programs were mostly for Openwave Up.Browsers, but we also needed another slightly different set of programs for Nokia cellular phones—kind of like programming specifically for Microsoft Internet Explorer and for Netscape browsers. We now had four different groups of programs for the presentation layer! Unfortunately, it got worse. Another version of WAP allowed for another language called WML (Wireless Markup Language), meaning that we had to write another group of presentation programs to handle this new language. The intermediate programs remained the same, and the database and data layouts remained the same, but the presentation programs changed.

In came XML (eXtensible Markup Language), the all-purpose, special presentation focused language. The idea was that since everyone was trying to write dozens of programs to satisfy this and that, why not just ensure that the information being passed between the presentation and logic layers was all the same? This meant that there would be very little presentation program changes, since all the parameters of the data would be the same. This is obviously a very simplistic view of XML, but the point is made.

Security

We've seen many tiers involved in data access from presentation to logic to data, with multiple data distribution methods and multiple client server configurations. It's enough to make your head spin in confusion! And whenever *our* heads start spinning, the first thing we think about is security. So much is happening, how can we ensure that we don't miss something? We surely don't want to suddenly find that our data has been stolen, lost, or corrupted.

These issues bring about security concerns such as:

- Data must be protected from specific users and/or unknown users.
- Users must be protected from other internal users and/or unknown external users.
- Enterprise access must be protected from internal users and/or external users.

Since each process (presentation, logic, and data) can implement its own set of business rules, access rules, and access points, security is a real concern. If you've ever lost data or found a virus on your computer, you've felt the pain of having a poor security system.

In the wireless world, security becomes a very real issue. Data can be downloaded onto a PDA unit and taken anywhere you can imagine. If for some reason that unit is lost or stolen, your data might end up in the wrong hands. Worse yet, if the data is of a personal nature to your clients, you might find yourself in a legal battle.

PDA Units

PDA units are excellent as presentation-layer front-end devices. The screens are small, but can still show quite a bit of information. They have processing power and memory and are therefore also able to manage the logic layer. Since these units have the capability to hold databases, large and small, the data layer can be contained on the individual units as well. In effect, a single Compaq iPaq could be considered a single-tier, all-in-one system, very effective for salespeople and portable applications.

Referring back to the client-server configuration diagram in Figure 3.3, all mentioned setups are possible with PDAs, especially with wireless modems as connection to the Internet allowing the dynamic extension of corporate enterprise solutions. Data can be distributed in a push or pull fashion, and with pass-through mode, on the PDA even the distributed presentation configuration is possible.

These units definitely open the possibilities of remote access to corporate data, especially legacy systems or data warehouses, which can really give the mobile end user a real advantage over competitors. Empowered by the entire corporate historical data (with probabilities predetermined) and client history (with trend and co-relation among bought and available product), cross-selling becomes a dynamic and very profitable potential for mobile workers.

Final Thoughts

PDAs are fast becoming a part of enterprise systems. As memory and real-time synchronization potential for these devices increase, they will become more active in our daily lives. As such their interaction with the existing systems will become more distributed. Since these units have strong processors, they will run the spectrum of tiered solutions and provide the users with many possible and viable solutions to corporate data, as well as Internet-based information. Who knows, maybe one day they'll be used as timesharing servers in a distributed environment.

CHAPTER 4

Data Warehousing

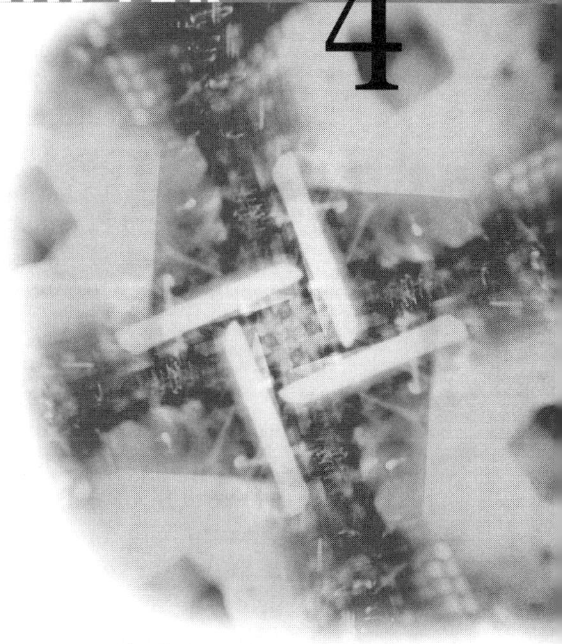

Chapter 2, "Database Architecture," spoke of databases and how they came to be, with a highlight of PDA considerations. Chapter 3, "Client-Server Architecture," spoke of client-servers with an emphasis on PDAs as an extension of the corporate enterprise. Since data is the corporate world's most valuable asset, many firms are centralizing their information into one main area called a *data warehouse*. By combining databases, client-server methodology, and a pool of corporate information from a data warehouse in a mobile wireless environment, the PDA becomes the next major portal to enterprise data.

Many firms find very valuable information in previously recorded data. If you know what a particular person's spending habits or purchases have been over the past several years, you could focus a specific marketing campaign toward that person if there was a history of purchasing a specific product. If Ralph has a history of purchasing a ski vacation package every winter, wouldn't it be a good idea to put Ralph at the top of your potential purchasers list if you're selling ski vacation packages? Why waste time on someone who never purchased one before? A data warehouse allows this type of analysis to resolve and/or focus on business concerns, patterns, and trends.

What Is a Data Warehouse?

A data warehouse is what it says—a warehouse of data. A warehouse is a place where items, be it shoes or medical supplies, are stored with expectations of being reclaimed or queried at a later time. In our case, the item being stored is data.

Data is stored in a warehouse so that whenever it is required, we can find it all in one place rather than having to gather the individual bits of data from all over to obtain the information we need. It's much faster accessing one central area than having to access a mainframe application, an NT server-based database, a flat file on a Unix box, and so on in the same instance. With a warehouse, all that we want is, generally, in one central area. If it is not, we can arrange it so that the data is in one place by starting a new project to load data from a specific source into the warehouse. These projects are usually quite the undertaking in large organizations, since data must be analyzed and synchronized between the different source systems, which is quite time-consuming. For example, a telephone number in one system might be different from a telephone number in another system, and we must then determine which is the correct one. This also involves ensuring that they are in the same format. Just look at your personal telephone organizer—do you have the area code for all numbers, or just for those of out-of-town numbers? If someone else were to rely on your organizer for telephone numbers, it would probably not be useful because some of the area codes would be missing. For that information to be useful to anyone else, all must be in the same format—area code, and then telephone number. The point is that everything we require is in the data warehouse repository, and must be in the same format and synchronized between all the systems. For our example, all telephone numbers would contain area codes.

In a typical warehouse, we could have shoes from this year, last year, and maybe from years ago. The same idea applies to a data warehouse. We might have data from today, yesterday, last week, a month ago, a year ago, 10 years ago, and so on. A data warehouse contains historical information, and we therefore call it a *data warehouse historical repository*. It is, however, important to set a base granularity throughout the warehouse data.

Granularity

Think of granularity as a measurement. We want to make sure that all measurements are of the same type. For example, if we say the granularity is gallons, we know we're always talking gallons, not liters, not barrels, or buckets. If all measurements are gallons, then don't expect to count the number of liters without extra conversion effort. The same concept applies to a data warehouse. If we capture and load data on a monthly basis, do not expect to run a query or perform a calculation to determine daily quantities. If we're talking monthly we can't get daily, but if we're keeping daily facts and figures in a daily granularity warehouse, then, of course, we can determine monthly results since we know how many days are in a month. We refer to rolling up facts in this manner as *aggregation*.

It's quite normal to count daily facts and figures to determine monthly totals in a daily granularity warehouse. The nice part about aggregation at a fine granularity level is that if a monthly number or figure is under scrutiny, we can easily show how we arrived at the number because we have the individual daily facts to back it up. The granularity of a warehouse is the lowest level of justifiable and attainable information in a data warehouse.

If we were to capture daily data directly from its source and were asked to justify a captured daily number within the warehouse, we couldn't because it wasn't derived from the warehouse data but from a source. Applying this to a monthly aggregated number, we could drill down to the individual daily numbers since we captured the data on a daily basis. If we were to justify those individual days, we'd have to refer back to the originating source system, which initially produced the number.

Organized Structures

Users needing to access data in the data warehouse historical repository could simply perform simple or complicated queries depending on their needs and obtain all the information they want. However, that can cause major problems from a performance point of view. Data warehouse queries usually run long scans for large amounts of data. Imagine a large number of people going into a shoe warehouse looking for specific shoes. Some people might be looking for women's blue shoes, others for men's size 11 shoes, and still others just want shoes that are on sale that day. It would be a mess with people rushing in and out, shoes not being returned to their respective boxes, people at each other's throats over the same pair of shoes, and so on. The best thing to do would be to ensure that everyone knows just what they are looking for and gets only what they want so they can be in and out of the warehouse as quickly as possible. This would allow for a structured and organized approach to shoe shopping and would free up the helpers so they could assist more individuals during the shopping hours.

In a data warehouse, we want to organize our data in the same way as we would the shoe warehouse so all users will get what they want, when they want it, and as quickly as they can get it. This, of course, implies that we have a pretty good idea of what the users want beforehand. If we do, in the shoe warehouse example, we could put all women's shoes in one corner and all men's shoes in the other corner, each ordered by shoe size. Then we could place all shoes on sale for that day by type on the front display racks. We'd then organize the shoe distribution according to anticipated wants. We're, in essence, controlling the flow of people and directing them to specific areas so they don't all trip over each other. In the data warehouse, we want to do the same thing. We can place customer information in one particular table or group of tables, account information in another group of tables, and so on. These are called *subject areas*. To take this one step further, we could copy specific data into their own physical areas of access. For example, we can organize and group together data on all customers who reside in Canada and their respective accounts. Of course, we'd only do this if we knew that someone was regularly looking for accounts of customers who reside in Canada. Obviously, it would go against reason to include customers who reside in Switzerland in this particular area because they wouldn't be required, so why include them?

A data warehouse is, essentially, an organized and structured central historical repository of data.

Architecture

Think of a data warehouse as a process that includes many pieces. While many people refer to the data warehouse historical repository as the data warehouse, it's important to understand that the data warehouse is the overall process that includes ETL steps (which we will define later), a historical repository, and usually one or more data marts.

Let's examine the components of Figure 4.1. We'll only slightly scratch the surface of each of the components to give a general introduction to the respective topics. There's a lot more to each of these areas, with many books available on each.

Source

On the left side of Figure 4.1 are the source systems. A data warehouse is usually the centralization of data from multiple legacy source systems. Figure 4.1 shows three legacy source systems: Retail Banking, ATM, and Mortgage Loans.

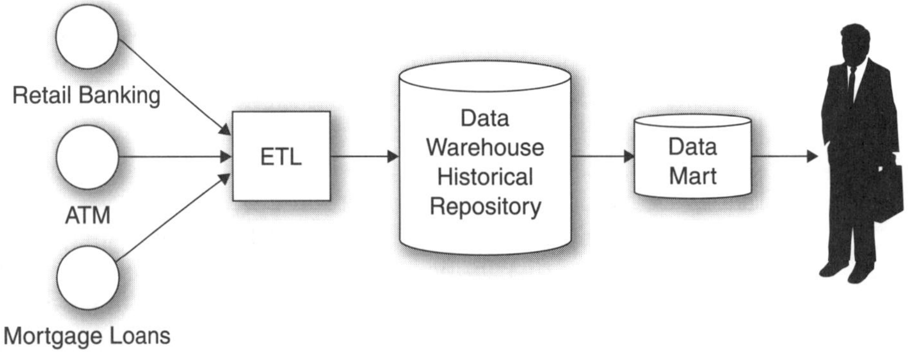

Figure 4.1 Data warehouse.

Data from each of these systems is captured and synchronized into one version of the truth. For example, the retail banking system might contain an address for a particular customer, while the mortgage loan application might contain the same customer but a different address. The problem we'd face is, which address is the correct address for the customer? As you can see a merging, comparison, and synchronization of information process must be performed. This will ensure that all who access the information from the data warehouse will have the same information, thereby allowing the end users from across the enterprise to all view the exact same information throughout history.

Note that we didn't mention an address for the ATM application. This was intentional, as not all applications will all have the same information. Applications might have their own unique information and might contain the same data as other systems, even thought it might have been gathered independently.

Of course, data in certain legacy systems might be incorrect, and this process will help in identifying these instances and allow the individual department applications to correct their information. This is an entirely different topic out of scope from our focus, but definitely a major side effect of data warehousing.

ETL

The data from each of these source systems is captured via what is known as the Extraction, Transformation, and Load (ETL) step. This ETL step represents many processes, which include the *extraction* of data from the application source systems (such as Retail Banking, ATM, and Mortgage Loans), the *transformation* of that data into a common format and structure, followed by a *loading* process, which loads the data into the data warehouse historical repository.

There is quite a bit more to the ETL process, but it suffices to say that this step is the most complicated aspect of a data warehouse. The developers must really understand all possible values of data being brought into the warehouse from each source system. Each piece of information must be verified, cleansed, and transformed into the repository's common model.

This process involves much cleansing of data, synchronization between multiple inputs, very much involves the business users from each source system to verify which value is truly the appropriate main source, and then a clear understanding of all aspects of that data from expected values to overall dependencies between the individual data values.

We think of this step as mining for diamonds. First we find the mineral, verify that it is indeed a potential diamond, clean off any unwanted material, transform it by cutting it into the desired shape, and finally placing it in the store (the repository).

Repository

The data warehouse historical repository holds all the data warehouse data. Not only does it contain the most recently loaded information, but also all previously loaded information at the predefined and architected granularity. It is very important to understand that information in the repository is of a historical nature. The whole point of a warehouse is to capture information from distinct systems, centralize it, and keep the information over time. Normal operational applications such as the retail banking system usually do not keep historical information. For example, if a customer changes her address, the old address is usually not kept in the system. Therefore, by loading the customer address into the data warehouse, we can keep a history of this customer's address over time—if we find this to be a valuable bit of information.

Since we're keeping data over time, there tends to be quite a large amount of data accumulating in the repository. For this reason, it's important to only store data that is required by the business user. Do not keep extra fields of data in the repository knowing that it won't be used in the near future and hoping that it might be requested in the future. This is a common mistake in a data warehouse environment. People hold on to information thinking that it will save them development time and effort in the future when the end user finally does decides that he or she really does want that one bit of information. You'll wonder where all your disk space went only to realize that all that extra unnecessary data gobbled it up over time. Take this hint: Only load exactly what is required and nothing more.

A repository is a database, and databases are modeled. The repository database can be modeled in several fashions, but the most common is called *third normal form* (3NF). Data modelers usually have certain rules to model data and the structures in which it resides. 3NF is a detailed method of recording business requirements at the data level. It also allows an open architecture, since each piece of data (name, account number, address, account balance) is individually stored in its own area—to a point. In this modeling methodology, redundancy is minimized, insert and loads are optimized since inserting into one area might not affect many other areas, and there is much focus on the overall data integrity.

Data Marts

Once the data warehouse historical repository is populated, the data is nearly ready for the business users. Remember that the end users specifically requested the data in the repository, with users requesting specific data for their specific needs and usage. Many source systems were accessed and used to populate the repository over many iterations of the same granularity. This means that the repository is full of data, but not necessarily all required by the same user. This is where data marts come into play.

A data mart is usually its own database, just as a data warehouse historical repository is usually its own database. The data mart is created from the data in the data warehouse historical repository and is, therefore, a subset of that historical data. You cannot have data in a data mart unless it's initially in the data warehouse historical repository.

Since data marts are a subset of data in the historical repository, they can therefore be recreated at any time. Remember that many users requested specific data for their own use. This, therefore, could be translated to a completely distinct subset of the repository, thus becoming its own data mart, consequently, a data warehouse can have multiple data marts. End user A could be using customer name and address from the Customer table along with account balance from the Account table, while end user B might be using customer name and telephone number from the Customer table but no account information at all. Each data mart is its own unique environment. To go back on what was said a moment ago, a data mart could also be a logical grouping of data and therefore reside within the data warehouse historical repository and not physically distinct. This is especially true for very large databases where duplicating information is very time consuming and disk space is a concern. Normally, however, a data mart is physically distinct in its own database environment.

Repositories are usually modeled in a 3NF design, while data marts are almost always modeled in a Star design. As you can see, the tables in Figure 4.2 are set up in a star-like fashion; hence the name Star model.

The idea behind a Star design is its simplicity. The models are built for nontechnical business users who simply want to query the data. Placing all customer data in one big customer table makes accessing the customer data quite simple. If you want customer name, check the customer table; if you want account information, check the account table—very simple and very intuitive. These specific tables are called *dimension tables*. There's always one extra special table called *Time*. Since the whole idea behind a data warehouse is to view historical data, we must have some method to determine if we're looking at last month's data or last year's data. This is accomplished via a table of dates, with the most common denominator being the data warehouse's basic granularity. If we're capturing information on a monthly basis, the lowest level of granularity in the Time dimension table will be months. No week or day entries will exist because the data is not recorded at that level.

The middle table in Figure 4.2 is quite interesting. In this example, it's called the monthly account balance fact table. Every data mart has at least one fact table. Fact tables usually contain numerical cumulative values at the warehouse

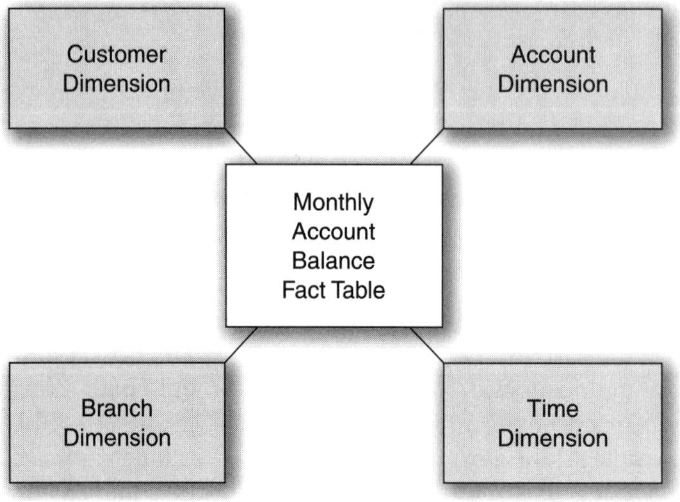

Figure 4.2 Star model.

granularity; hence, in our example, the word *monthly* in the table name. Each entry in our example table is per Customer's individual Account by Branch per month. The idea is that the business user can add the individual monthly account balances by any of the dimensions associated with the fact table. This really allows end users the freedom to aggregate the individual account balances by whichever dimension they want, which is a very flexible design and highly useful for the business user.

Because we're trying to place all information for each subject area into its own basic dimension table—for example, all customer information into the Customer table—we're increasing redundancy. If customer Ralph has an address change, we'd simply add an entirely new row of data with the only thing changing being the address. All other information would remain the same. If you were to query the table looking for Ralph, you'd see two entries instead of one as you would in a 3NF model. This means that the end user must be more precise in his requirements and perhaps re-query the data looking for all customers in a particular area, thereby resulting in only one result row for Ralph. Alternatively, maybe run the query with a specific time period, which might reduce the number of Ralph rows in the final result—get the idea?

This topic can span volumes, but unfortunately, it is not the point of this chapter. What we're trying to show here is how to take large amounts of data from the data warehouse and place it on a PDA. First, however, we had to highlight the major components of the data warehouse. Next, we'll look at the PDA extension portion.

PDA Data Warehousing

There are many reasons why organizations want to build a warehouse. Those reasons can be to centralize information from multiple application systems and get a single view of the data for the enterprise, retain historical data, use the data for cross-selling purposes from existing clients, make trend estimations based on past performance, and so on. The belief is that the enterprise could benefit from the existing data, so don't simply overwrite data and lose it forever; keep it around in an organized and structured historical repository for specific strategic analysis.

With the arrival of portable and mobile wireless systems, organizations are now looking at offering their information on PDAs. Employees would then have access to a wide range of information anytime, anywhere, giving them access to a plethora of past information.

For example, if a salesperson could download his firm's available inventory and his specific client's past purchases, he could certainly benefit from a projected purchasing trend analysis as well as promising product shipping and delivery based on goods on hand.

Data Stores

Data warehouse historical repositories are known to contain quite a bit of information. Data marts are subsets of the repository usually built using a different type of design that contain much redundancy in the dimensions and therefore potentially large fact tables as each instance of each dimension table pertaining to the fact table might contribute to a new row. Over time, this will certainly add up to lots of rows and, therefore, a lot of disk space.

One thing the PDAs of today do not have is a lot of disk space. Most people have only 32MB of extra memory on their devices. Although memory extension cards are being developed to increase capacity to 5GB of information, most users do not have this feature. Personally, I have 64MB on a memory flash card, nice for personal notes, addresses, and e-mail, but not really much from a data warehouse point of view.

The point is that with little disk space (read "memory space") available on the devices, data saved to the devices must be planned very carefully. Data marts are designed for specific usage defined by the business users, and they also give the users the overall capability to query data on an ad hoc basis. Information on PDAs must be much more specific and tailored to specific business uses. Perhaps as the memory capacity on these units increases, users will have the luxury of loading the devices up with GBs of data to use and query as they wish; however, at the moment, every bit of data must be specifically architected.

Databases such as Sybase iAnywhere's UltraLite solution require each database query to be predefined before deployment to the PDA. With this methodology, only the columns and rows required by the PDA application will be populated. Even with this reduction in data overhead, we might still have too much data to upload. We'd then turn to business rules to further dictate usage. The individual commands of the PDA application might query a table, but the uploading can be further optimized to only capture rows pertinent to the individual. For example, the PDA application might look at all customer names, but the data upload process might only load specific customers to limit the amount of data on the device. Every byte of data uploaded to the device must be thought out in the design and architecture of the application.

The final data store for the PDA is small, very specific, and yet still pulled from the same data warehouse historical repository.

The Vineyards application in Chapter 10, "Sybase," uses the data warehouse strategy introduced in this section. The Sales History option is very specific and shows a predefined usage of data in the application. The example shows the quantity of wine ordered per month over the past 12 months for each type of wine per customer. Normally, in a data warehouse, we'd simply aggregate the results from a base fact table as the user requested them, but presummarizing the values for specific customers and distributing this table to the PDA unit is a big memory-saving strategy and the concept behind data stores.

PDA Source

Another point to mention, which should be obvious from a data warehouse point of view, is that all data from the repository is read only. Data loaded into the data warehouse historical repository must be captured, cleansed, and transformed in the ETL step and at no other point. All feeds to the data warehouse must be at the onset of the ETL process.

If a mobile user captures new data such as sales on the PDA unit itself, it should be in a completely distinct table. The table and information would be downloaded to the ETL machine at synchronization time and, at the warehouse granularity interval, the data would be processed and loaded into the historical repository, as would all other information from all other source systems.

If data were loaded into the repository on a daily basis, the captured information on the PDA would be downloaded to the server daily and allowed to enter into the input ETL data warehouse process. If the data warehouse granularity were monthly, all PDA information captured would be integrated into the data warehouse ETL process monthly. However, some common sense is required at this point. We would not want to retain all the sales information on the PDA device for 30 days, as the risk of losing it is too high. The architecture would most likely require all salespeople to download their sales data daily or every two to three days to ensure that these orders and statistics are not lost. The data would then be captured on the ETL server and saved until the monthly load process begins. Only at this point would all the past month's data run through the ETL steps for loading into the data warehouse historical repository.

PDA Empowerment

Organizations have been building data warehouses for a number of years and are now moving toward merging their entire individual warehouses into enterprise data warehouses (EDW). With this move comes the request from a multitude of diverse users throughout the organization to access the information from different portals. Mobile devices are one method to accomplish this connectivity and to empower the users with historical information access on their personal digital assistants.

Final Thoughts

PDA applications, as with any other type of application, require data. Data can be either accumulated at the PDA or brought over from a server. Data warehousing is read-only data, thus it would have to be uploaded from the corporate server. This information would allow the mobile business user access to historical information and, therefore, potential trend analysis or other potential uses.

Since PDAs have little memory and data warehouses usually have large quantities of information, there would obviously be space issues and, consequently, design issues. Developers must take into account the use of the information, and, therefore, exactly which information is required. Moreover, rather than aggregating information on-the-fly on the PDA, it might be best, and it is recommended, to pre-aggregate all data before uploading to the mobile units. This will save time as less processing is required on the device, occupy less memory, and ensure that the business requirements are fully analyzed.

If data were to be captured on the PDA, it would not be part of the data warehouse until downloaded and merged with the warehouse through the ETL processes and then re-aggregated and uploaded to the PDA. Any information gathered by the PDA is considered a new source of information and should follow all legacy source system methodologies.

CHAPTER 5

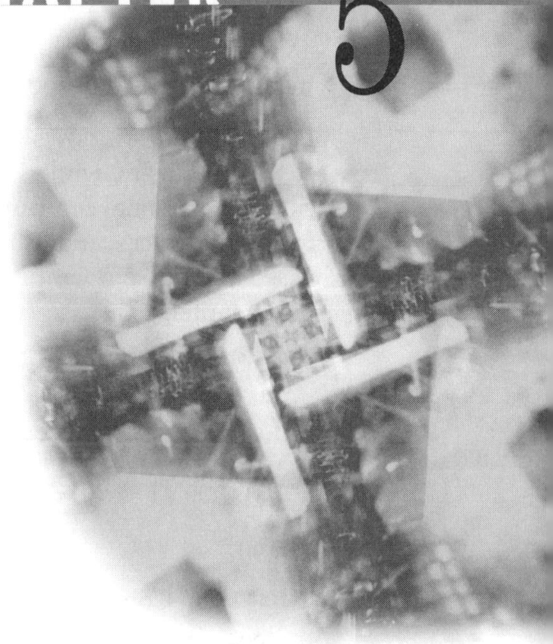

Palm

There are basically two types of wireless devices available in the marketplace. By far, the most popular devices are those from Palm. To be more precise, we should say devices from 3Com, since they bought U.S. Robotics back in 1997, who in turn bought Palm in 1995. Palm was founded in 1992. However, everyone says "Palm," and so shall we. According to IDC figures for 2001, Palm was the leading global provider of handheld computers with a 41.5-percent share of the worldwide personal companion handheld device market, and a 60.3-percent share of the worldwide handheld OS market. Palm products are sold in more than 54 countries and through Internet retail Web sites.

Palm has nearly 200,000 registered developers, more than 13,000 software applications, and more than 100 add-on devices. Palm handhelds are growing increasingly pervasive as information management becomes ever more mobile. Palm believes that handheld computing is the next wave in individual productivity tools for the global workforce just as we do. This growth of the handheld computing market is, in part, being driven by the transformation of the corporate environment into an extended, virtual enterprise supported by a highly mobile, geographically dispersed workforce requiring fast, easy remote access to networked resources and electronic communications. Palm continues to provide solutions to address this growing demand with products that are simple, wearable, and connected.

The second most common devices, in our opinion, are Windows CE (a.k.a. WinCE) based. Microsoft's Pocket PCs (Windows CE based) are advancing at an incredible rate, Palm still holds the majority of the market, but we believe that Pocket PCs will soon dominate.

Palm OS and Windows CE are completely different. Palm is focused on personal organizers, while Pocket PCs are just that, a small, trimmed-down version of a personal computer. Pocket PCs have the Windows operating system for small devices, and as the name implies is based on the personal computer-based Windows operating system. We will focus more on Pocket PCs in the next chapter.

History

Here is a brief outline of Palm's history from its early years in 1992 to 2002.

- 1992
  - Palm Computing is founded.
- 1996
  - Mar Pilot1000 and Pilot5000 organizers are introduced.
  - Jun First international launch, in France.
  - Fall *PC Computing*'s MVP "Usability Achievement of the Year" award goes to Pilot 1000.
- 1997
  - Mar
    - PalmPilot Professional and Personal Edition models are introduced.
    - There are a total of 2000 registered developers
  - May 3Com purchases U.S. Robotics who own Palm Computing
  - Sep IBM's Workpad introduced (based on Palm OS).
  - Dec Palm begins licensing its Palm OS platform.
- 1998
  - Feb Palm launches first eNewsletter.
  - Mar Palm III is introduced.
  - Sep
    - Sybase forms alliance.
    - Handspring licenses Palm OS.
  - Dec 3595 registered developers.

- 1999
  - Feb — Palm IIIx and Palm V handhelds introduced.
  - May
    - Nearly 73% of US market and 68% of world markets for handhelds.
    - Palm VII introduced.
  - Jun
    - TRGpro introduced—first Palm OS device with expandable architecture.
    - 13,751 registered developers.
  - Jul — Palm IIIe introduced.
  - Oct
    - Palm Vx and Palm IIIe Special Edition introduced.
    - Nokia licenses Palm OS platform.
  - Dec — Six-millionth U.S. patent awarded to Palm HotSync technology.
- 2000
  - Feb
    - Palm IIIc color and Palm IIIxe expandable handhelds introduced.
    - Palm Portable Keyboard introduced.
  - Mar — 41,000 registered developers.
  - Apr — 65,000 registered developers.
  - May
    - Palm HotSync Server software introduced.
    - Palm Ethernet Cradle introduced.
    - AnyDay.com acquired.
  - Jun — Actual Software acquired.
  - Aug
    - Palm VIIx introduced.
    - Palm m100 introduced.
  - Sep — 100,000 registered developers.

- Nov
    - Palm and IBM partner.
    - Palm Mobile Internet Kit ships (links handhelds with mobile phones).
- Dec Day-Timers and Palm partner.

- 2001
  - Jan
    - Palm and Sprint partner for CDMA wireless Internet access.
    - 140,000 registered developers.
  - Feb Palm goes Arabic.
  - Mar
    - Palm m105 introduced.
    - Palm m500 and m505 introduced.
  - May
    - 10,000th Palm OS application.
    - Palm launches in India.
  - Jun Bluetooth card introduced.
  - Aug PowerPoint files available on Handhelds.
  - Sep Palm m125 introduced.
- 2002
  - Jan
    - 20-million devices sold.
    - Palm i705 introduced.

Overview

A recent interview with Palm's vice president of marketing really explained Palm's vision quite well. Palm units are built as personal organizers. The devices are indeed microcomputers, but they are focused on the active professional. They in no way focus on being a smaller version of a personal computer such as a Windows or Macintosh computer.

Many people refer to the Palm units as PalmPilots, but that's incorrect. Initially, the first units sold were referred to as Pilots. It's only when these early units became popular that Palm decided to rename them PalmPilots just so there'd be no confusion with the Pilot pens, which were also quite popular at the time. After a while the name was again changed to Palm and Roman numerals were appended, with the first units being called Palm III. We've all heard about the Palm V, Palm VII, and so on (just out of curiosity, whatever happened to Palm IV and VI?). Today there are many types of units as seen in the next section. Throughout this book, we'll be using the Palm m500 (Figure 5.1).

With all the different units out there, 3Com has begun referring to the Palm Computing Platform rather than the individual units. This brings the Palm operating system to the forefront for future possibilities.

There are many types of Palms, many special characteristics to the Palm OS, and of course the capability to connect to other devices, most notably to the desktop personal computer. This is about the only way Palm has any connection with Microsoft since, as you well know, the majority of desktops in the world are Windows based.

Before we miss a major point, we should point out that there are many (and we mean many) more applications written for the Palm OS units than for the Pocket PC units. Palm has been around a lot longer than Pocket PC, and therefore it stands to reason that there are many more applications out there. The Palm Web site boasts about the number of registered applications—the actual

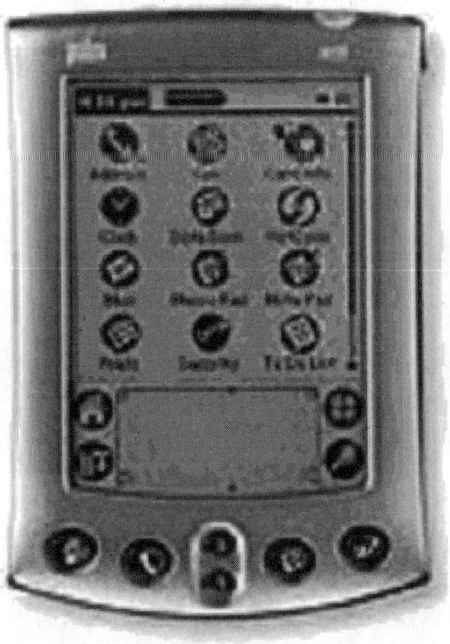

Figure 5.1 Palm m500.

number is anyone's guess, but let's just say that there are thousands and thousands of programs. Many are free and some are not. Some are focused on the individual, and others are more corporate based. The point is that the platform is open, which has allowed thousands of developers to design new programs over the years.

You can try adding a variety of applications and functions to the Palm, and the best place to find these is on the Palm Web site at www.palm.com. From there, you'll be able to find lots of other popular sites and applications.

Physical Units

There are many different types of units and not all are from Palm. One of the reasons Palm is referring to their operating system more and more is to bring attention to the point that Palm OS doesn't only run on Palm units. The Visor, for example, also uses the Palm operating system.

Devices

What are the different types of typical Palm devices? You can find a listing of them at www.palm.com/products. When selecting a device, be sure you know what options you need. It's the same as with shopping for a car—you wouldn't buy a Volkswagen Bug if you were going off-roading, and you wouldn't buy a Ferrari to do errands and drive the kids to school—well, most people wouldn't!

Of course, there are other units available from other vendors such as TRG, HandEra, and Sony Clie, to mention a few. We don't want to limit your view of Palm OS devices, but since this is a Palm-focused chapter, we figured we'd only show Palm units.

Cradles

All the current devices come with a cradle. A cradle is the base station a unit sits in while being connected to the desktop. Cradles have HotSync buttons—one press and with the software on the desktop, information is uploaded, downloaded, and synchronized.

While in the cradle, the handheld is recharged (for those units with rechargeable batteries). The unit just slides on top of the cradle and all the connections are automatically made if the cradle is connected to the desktop and the desktop is powered on.

If traveling and you don't want to carry this bulky cradle, Palm has a HotSync cable too. Simply plug the cable into your USB port and connectivity is fully available. On the downside, if using the cable, no battery recharging is possible.

For those who don't have a USB port, both the cradle and cable come in serial connection versions.

M500 Unit Specs

Let's have a closer look at an m500 unit. Over the generations, the units have changed considerably, but the latest ones have roughly the same features externally. Figure 5.2, from the Palm documentation, shows the basic components of a typical Palm device.

The m500 unit has an expansion card slot. Simply slide in a Flash memory card to expand the unit's physical memory capacity. Cards are available in 32, 64,

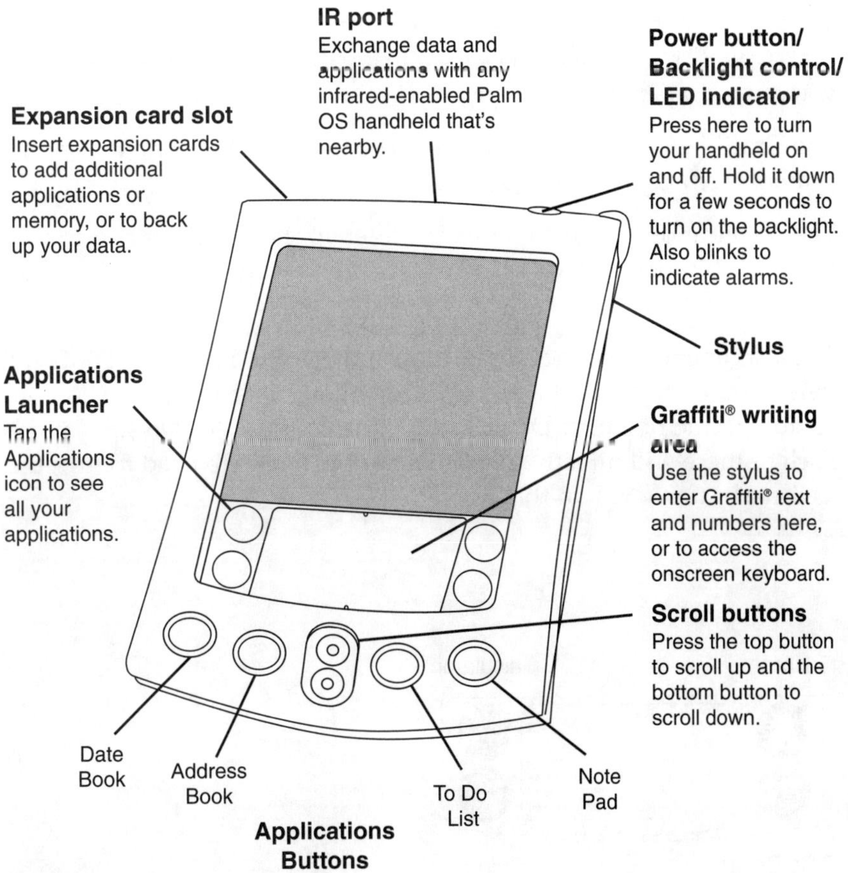

Figure 5.2 Palm m500 unit.

and 128MB. Software is also sold on these memory cards, and installation is usually quite simple.

The infrared (IR) port allows units to communicate wirelessly. We've been to several business functions when a person has asked us to beam our business card information to their unit—very useful functionality. We also use the IR port to beam files from a m500 unit directly to a laptop. Just point the two units together and beam a file. The other unit will instantly recognize an incoming file and ask for continuation permission—again, a very useful option.

The Power button is self-explanatory. It's only one method of turning on the unit. However, to turn on the unit and jump directly to a specific feature such as the Date Book, you can simply press the Date Book application button and *voilà!*—instant access to that function. By holding down the Power button for several seconds the backlight will come on. Palm devices are designed for a persistent state. This means that when you turn it off and back on (immediately or later), the unit will jump to where you were the last time.

To the right of the Power button in Figure 5.2 is the stylus, and we all know what that is, right? It's the small stick used to tap the screen more efficiently. Of course, you can use a pen or your finger, but the stylus will extend the life of the screen.

The Graffiti writing area is used to write freehand by using the special Graffiti symbols we're all so used to now.

The Applications Launcher area is preset applications defined to the soft buttons. Soft buttons, shown in Figure 5.3, are the four icons on the screen next to the Graffiti area.

The last functions are the Application and Scrolling buttons. The Scrolling buttons are used to scroll if an application allows it. The Application buttons are set to specific defaults: Date Book, Address Book, To Do List, and Notepad. Each of these will directly invoke its application—no need to turn the unit on, as these buttons will do that as well.

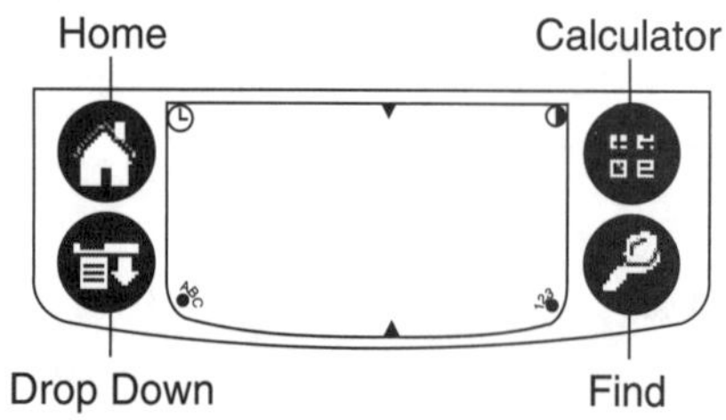

Figure 5.3 Applications Launcher soft buttons.

So, what happens when everything goes wrong? Just reset the device. There are two types of resets: hard and soft. For a soft reset, you have to use the stylus that comes with the unit. Just unscrew the top of the stylus and use the reset tip inside it to press the Reset button on the upper back of the unit, shown in Figure 5.4. This will stop whatever is happening on the unit and basically restart it.

A hard reset will return the unit to the original factory settings. Beware—this will erase everything you added to the device since the last hard reset or since the first time you took it out of the box. To perform the hard reset, simply hold down the Power button and gently press the Reset button with the special stylus device or a paper clip. Wait for the Palm Logo to appear and then release the Power button. A message will appear warning you that you are about to perform a hard reset—if you want to proceed, press the upper scrolling button; otherwise, press any other button to perform a soft reset.

Graffiti

Since a very big part of Palm computing is its special graffiti language, let's have a brief look at it. Graffiti is the special alphabet you can use for entering information into your Palm unit. These are special characters you draw with the stylus to type in words—if you choose to use freehand. These are unique to Palm and if you use them enough, you'll end up writing the Palm "shorthand" even when you don't have a device with you.

Figure 5.5 shows what each character looks like. Where you see the dark spot for each letter in the figure is where you place the stylus to begin the letter or

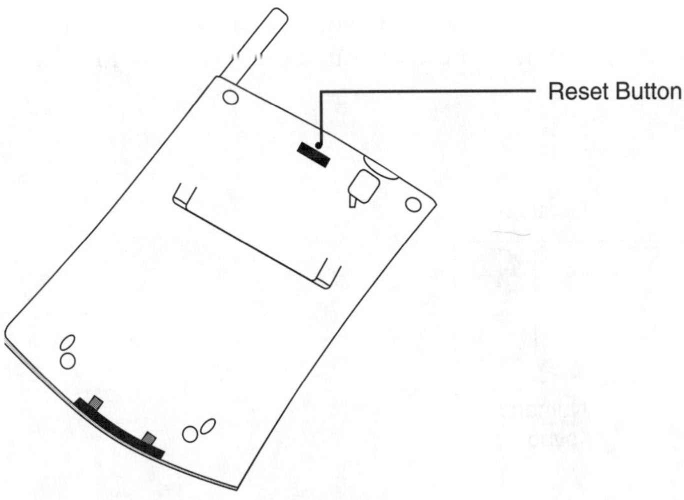

Figure 5.4 Palm m500 Reset button.

Figure 5.5 Graffiti area functions.

number. Yes, some look very odd, but they are supposed to be like that—when you start using Graffiti, you'll love it!!

If you look closely at the Graffiti area, you'll see four little icons in the corners. Figure 5.6 explains what these are for, which you'll find out by just by tapping them. Top left is a picture of a tiny clock. Tap there to invoke the date/time program. To the top right, is the half-white and half-black icon, which is the contrast application. On the bottom right is a 123 icon. Tapping this will invoke the keyboard function for numerical characters and special characters. You'll also see a tab for plain alpha and another tab for international characters. The bottom left is an abc icon, which invokes the same keyboard functionality as the 123 icon but with focus on the alpha keyboard.

Notice the little upward and downward pointers in the somewhat middle of the Graffiti area. These represent the writing division. To the left side closer to the Clock and Alpha Keyboard is where you'd write letters, and to the right of the division markers is where you write numbers. There is a lot to know about such a little area.

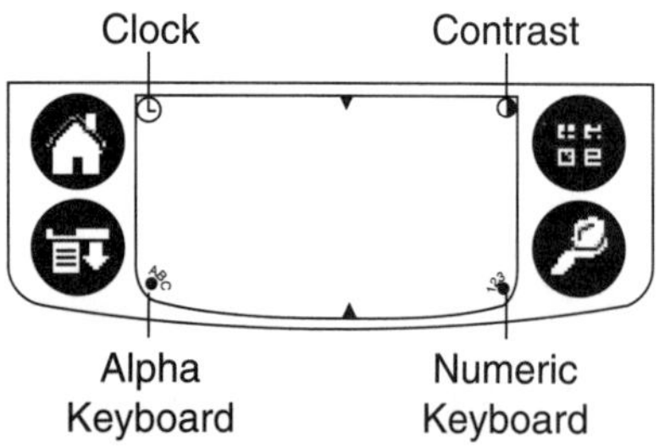

Figure 5.6 Graffiti area functions.

Palm Desktop

Now that we're all familiar with the units, let's look at the Palm desktop 4.0.1 and see how we can connect the device with our desktop. Palm units are great, but the main advantage to these devices is the capability to synchronize schedules, calendars, notes, email, applications, and the like between the main workstation, being our desktop personal computer, and the wireless device.

Step 1 in performing this synchronization is to pop the Palm Desktop Software CD that comes with the m500 Handheld into your CD drive. The CD's main executable will automatically begin and display the first screen. Select Install and the Palm InstallShield Wizard, shown in Figure 5.7, will start up.

Click on Next to jump to the next window. This setup screen will determine where the Palm directory will be located. Choose the default as shown in Figure 5.8, and click Next to continue.

Next, the wizard will automatically determine if Microsoft Outlook is installed on your machine. Since most of the world has Microsoft Outlook, the wizard will give you the choice of whether you want to pick Microsoft Outlook or the Palm Desktop application as your default desktop application. We highly suggest that you use Outlook for two main reasons. First, it's probably already installed on your machine and you're probably already using it regularly—so why change? Second, Palm works just fine with Outlook, so why not go for it!

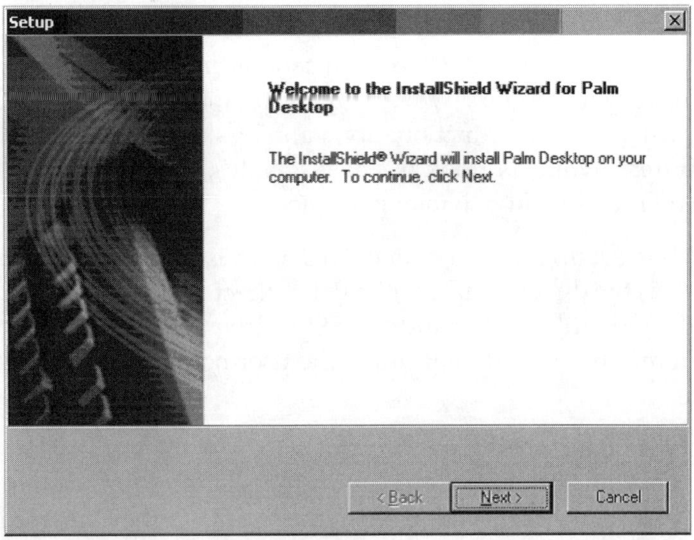

Figure 5.7 Palm InstallShield Wizard.

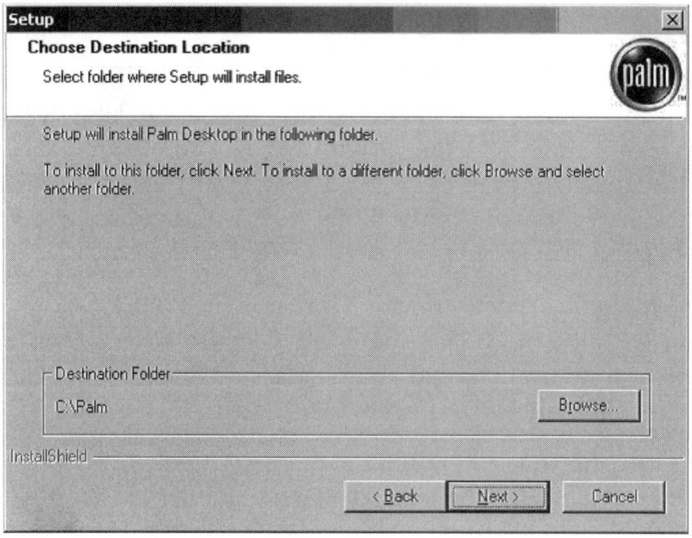

Figure 5.8 Choose Destination Location.

What's actually happening here is that the Palm InstallShield Wizard is setting up the conduit to access one or the other, but not both.

A conduit is the pipeline between the device and the desktop. It establishes a connection and translation between the PDA and the personal computer in order to exchange data. Without it, there can be no connection between the two machines.

Palm Desktop and Microsoft Outlook applications are pictured in Figures 5.9 and 5.10, respectively. Again, we highly recommend Microsoft Outlook. Notice how each has similar functionality; Palm Desktop has To Do, Notepad, Date, and Address, while Outlook has Tasks, Notes, Calendar, and Contacts. Both have good functionality, so take your pick.

The next step is to set up the User Name. This name is used for several reasons. One is to serve as the directory under the Palm directory (e.g., C:/Palm/Bob1). It's also used to identify your handheld. Since the software can be used to synchronize a number of different units, the user name will define the individual device.

Figure 5.11 shows the Create User Account screen of the installation wizard. Click the Next button and the wizard begins to install the application. Be sure that Microsoft Word or Excel is not open at this point—if they are, the installation will fail.

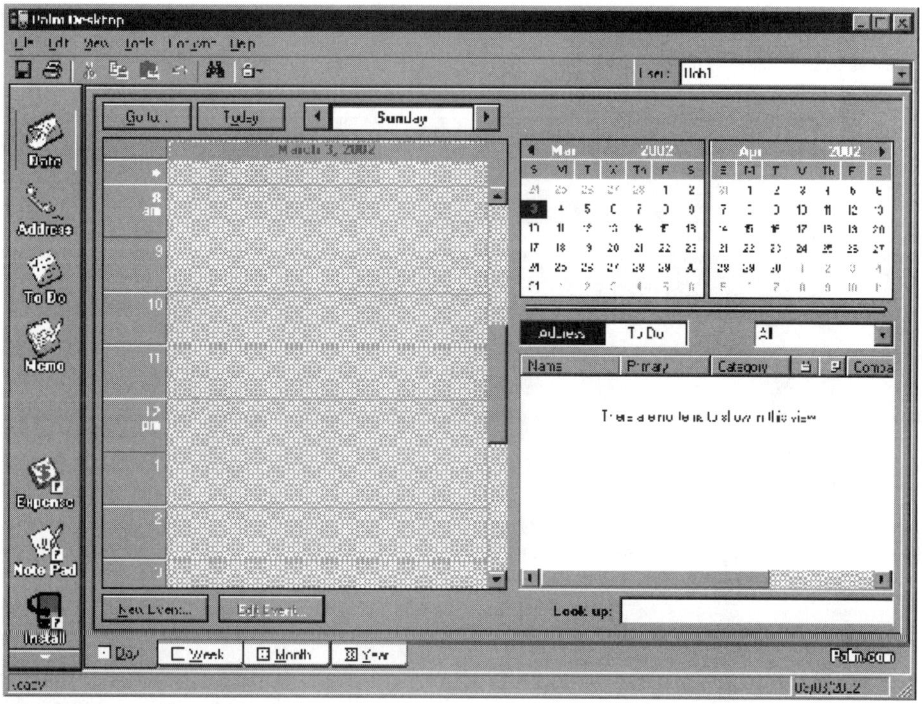

Figure 5.9 Palm Desktop.

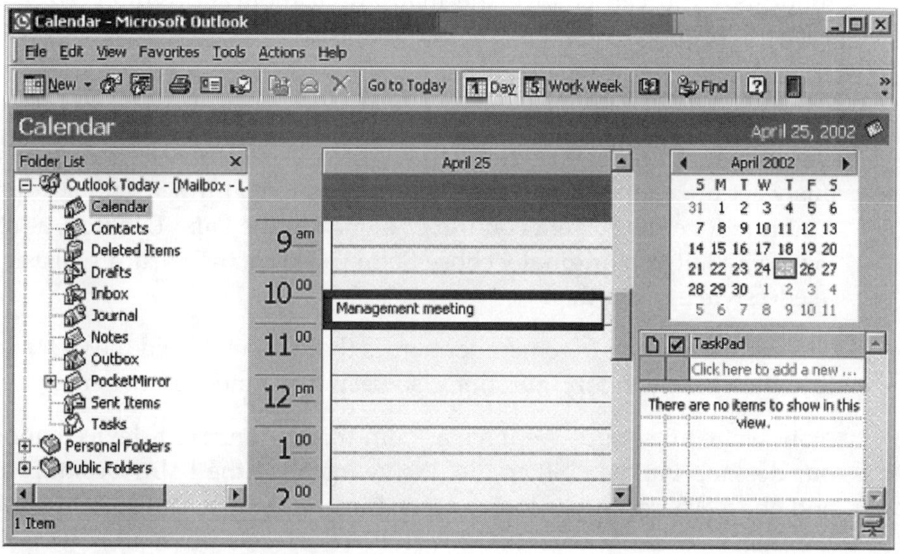

Figure 5.10 Microsoft Outlook.

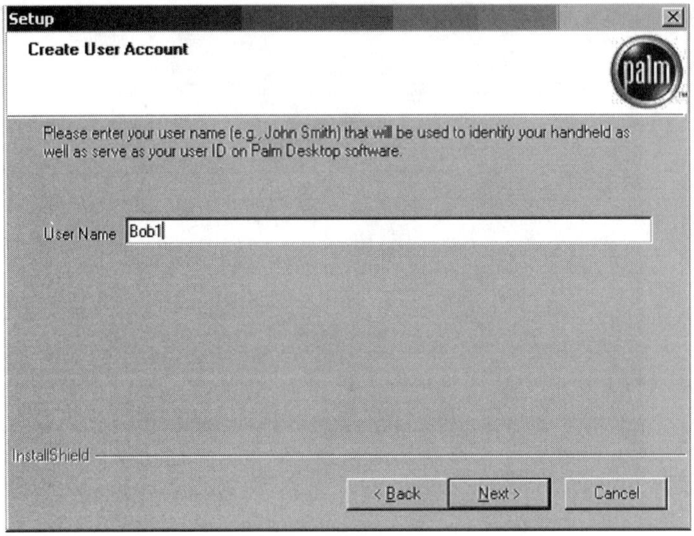

Figure 5.11 Palm InstallShield Wizard—Create User Account.

The last part of the wizard is to build a mail conduit so you can synchronize email on your desktop and on your Palm device. You can set this up now or leave it for another time.

If you decide to perform the mail setup now, select the mail application you are using on your desktop from the pull-down menu shown in Figure 5.12, and click Next. The wizard will begin to install your email synchronization conduits. You can change the mail application at anytime by clicking on >Start >Programs >Palm Desktop >Mail Setup.

If, at some point, you'd like to change from a third-party conduit such as Microsoft Outlook to the Palm Desktop, just use the CD to reinstall the Palm Desktop application. This time, the wizard will recognize that you've previously set up a conduit and will require you to confirm whether you want to continue with the current conduit or switch to the Palm Desktop, as shown in Figure 5.13. If you originally chose Palm Desktop, the wizard will ask if you'd like to switch to another.

An important point to remember here is that you will need the CD if you want to switch HotSync configurations, so keep it around somewhere.

Now we're all set up. We have a Palm m500 Handheld device connected to our desktop via the USB cradle. We've installed the Palm Desktop software, and chose Microsoft Outlook as our third-party synchronization application. We then selected Microsoft Outlook 97 or later as the mail server, since our desktop has Microsoft Outlook 2000. What's next?

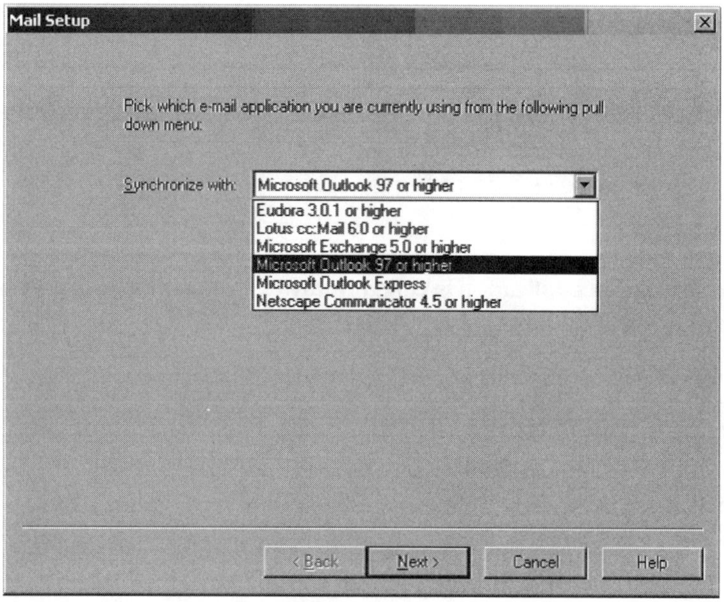

Figure 5.12 Mail Setup Synchronization.

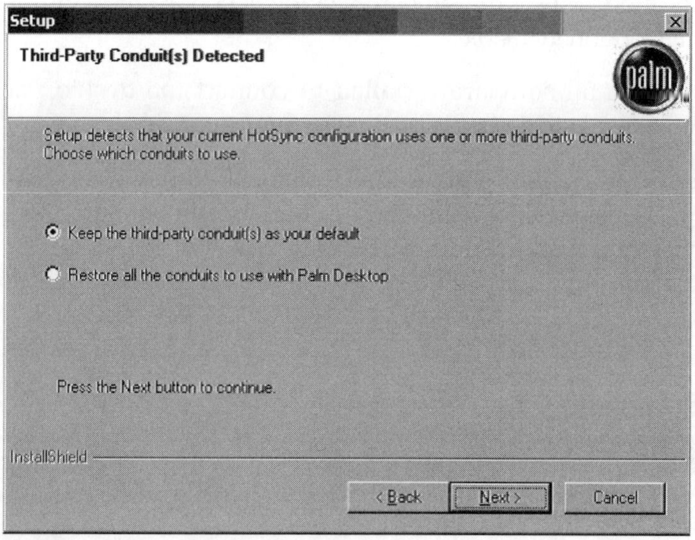

Figure 5.13 Conduit detection.

HotSync

If you look at your Taskbar to the right-hand side (assuming your Taskbar is on the bottom of your screen), you'll see a small round red/blue Palm Desktop icon for HotSync as shown in Figure 5.14.

NOTE

Palm's synchronization application is called HotSync. Microsoft's Pocket PC synchronization application is called Microsoft ActiveSync.

Click on this icon to bring up the HotSync functions, and then click Custom. The screen, pictured in Figure 5.15, will appear. This application allows you to select the HotSync synchronization actions per conduit. This is a very important application, as you can change the synchronization actions per conduit as you wish.

Select a conduit from the menu, and click the Change button. Usually, the screen pictured in Figure 5.16 will appear. As you can see, the HotSync action for Mail, as for most conduits, is to synchronize the files on both the desktop and the Palm device, have the desktop overwrite all mail on the handheld, or to do nothing. Each of these actions is temporary and will last until the next HotSync synchronization. Once completed, settings will return to the default, which is to synchronize the files—unless you check the *Set as default* option.

For the Mail conduit, you can also redo the Mail Setup to use another mail application. Not all conduits are changeable. If you have an in-house written application, the application will set the synchronization action, so the conduit will not have the change option available.

Now that we have all software installed to connect the desktop and Palm device, the next step is to actually do the HotSync synchronization.

From the Palm cradle, simply press the HotSync button. It looks just like the HotSync icon on the desktop. This will automatically activate the synchronization between the desktop and the Palm device.

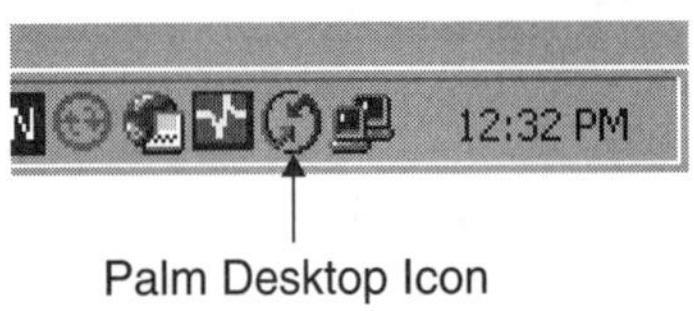

Figure 5.14 HotSync Desktop icon.

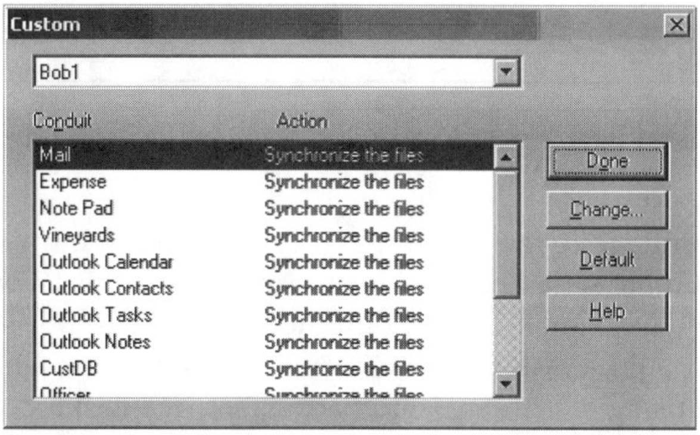

Figure 5.15 HotSync conduits.

The very first time you synchronize your device, you'll have to enter a username in the New User dialog box. This name must be the same as the one you selected in the Palm InstallShield Wizard under the Create User Account screen. Every unit should have its own unique name.

Once the button has been pressed, the HotSync process will run all the conduit HotSync synchronization actions selected. When completed, the Palm device will show a message stating the completion. You're done—the Palm unit is now synchronized with your desktop and you can safely remove the device from its cradle and be wirelessly mobile.

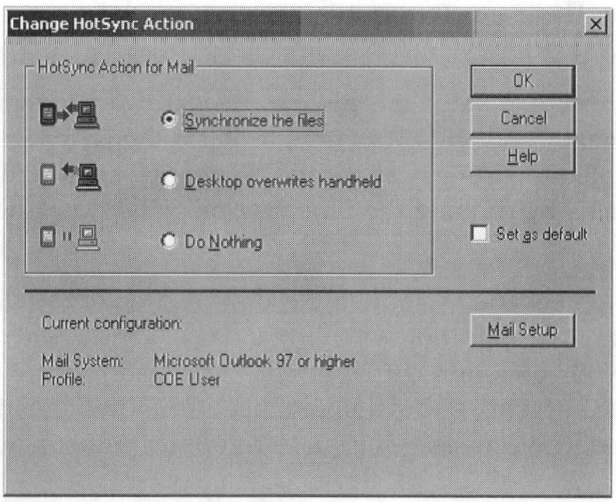

Figure 5.16 HotSync actions.

Beaming

Now let's quickly look at beaming information between your Palm unit and the Desktop or another Palm unit.

The m500 Handheld unit we're focusing on in this book has an Infra-Red (IR) port. This allows the unit to send and receive information via this IR port with other Palm units or the Desktop. Simply point the two unit's IR ports toward each other and they should register a connection—it's as simple as that, right out of the box. We're using a Toshiba ThinkPad T22 as our desktop and one day we sat the Palm m500 Handheld next to the laptop and wouldn't you know it, we heard a beep and a screen came on asking if we wanted to transfer a file. It is the simplest user interface you've ever seen!

To beam an application from the Palm unit, simply tap Menu, tap App, and tap Beam. Select either handheld or card, and tap the application you want to transfer. Tap Beam, when the Beam Status dialog box appears, simply point it to the receiving unit, and wait until the Status box confirms the transfer. To receive, point the Palm at the sending unit and follow the simple instructions. Again, it's a simple user interface.

Palm Emulator

There are two methods you can use to create and test Palm applications. The first is to write a program and upload it to the device, and the second is to use an emulator directly on the desktop. Both have their advantages and disadvantages. The following is a quick introduction to the Palm Emulator taken from Palm's Web site. There are other emulators in the marketplace, but we'll only look at the software from Palm.

The Palm OS Emulator is software that emulates the hardware of the various models of Palm-powered handhelds. An example of it is shown in Figure 5.17. It is extremely valuable for writing, testing, and debugging applications. Create "virtual" handhelds by running the Emulator on Windows, Mac OS, or Unix computers.

Emulated devices can be loaded with new Read Only Memory (ROM) software, so you can test your application with virtual devices, including different devices other than your own, foreign language ROMs, and debug-enabled ROMs for extra error checking and debugging features. The Emulator software does not include ROM images—it is like a computer without an operating system.

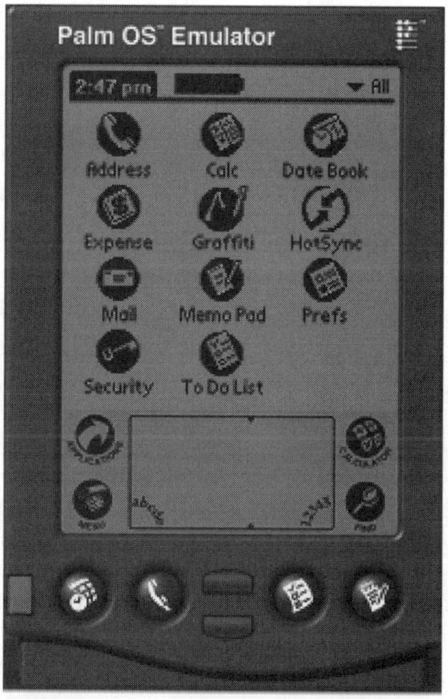

Figure 5.17 Palm OS Emulator.

To emulate a specific device you will need to obtain a ROM image that is compatible with that device. Typically, the ROM and device must match in processor type and display color depth. In order to use some device-specific capabilities, specific libraries and/or applications might need to be present. For example, to simulate Palm VII with a wireless connection you need the Web Clipping components (usually provided as part of the ROM). As new versions of the OS come out, Palm releases new ROM images that are set up to emulate older devices that have been updated to the new OS. When you pick a ROM image, the Emulator will give you the option to pick the device you want to emulate from the devices that could use that ROM.

There are three sources of ROM images: ROM image files downloaded from the Palm Resource Pavilion, licensee ROM images obtained via Palm's licensee's developer program, and ROM images downloaded from an actual device.

One last note on device emulation is that in order to make the Emulator cosmetically look like the device you are emulating, you will need to download and install the appropriate "skins." It is not necessary to use any specific skin to accurately emulate that device.

There are also two types of ROM image files: debug and non-debug versions. ROMs transferred from a device (and non-debug ROM image files) are designed to cover up errors. While appropriate to convince users that applications work, they interfere with developing truly reliable software applications. Debug ROMs are instrumented to reveal programming errors and techniques that might not work in future versions of Palm OS. Developers are therefore strongly encouraged to use debug ROMs to develop trouble-free software. Note that when a developer sees a problem using a debug ROM but not a device-transferred ROM, this is nearly always a sign that the extra instrumentation in the debug ROM is working. This is not a sign of a defect with the debug ROM! Developers should find their code, which triggers the reports, and make the appropriate changes. This also applies to the debug checks that can be turned on and off from the Emulator preferences. In general, you should test with all debug checks on and test anything that is flagged as an error. Turning off the debug check to make the error go away is not the correct answer.

If you don't have access to the ROMs yet and you'd like to start debugging right away, you can upload a ROM from most Palm-powered handhelds into the Emulator. Transferring the ROM requires a serial connection rather than a USB connection, and requires that nothing is installed on the device that has altered the ROM, such as TRG Flashpro. Do not upload device ROMs as a substitute for signing up for the Palm OS Developer Program and Development Seeding Program, or download the ROMs from the Resource Pavilion. You'll lose out on the advantages of testing your applications on a wide variety of devices, and you won't be able to take advantage of full testing capabilities without debug ROMs from the Resource Pavilion. Moreover, regardless of whether the development tool you choose comes with the Emulator, you should always download and use the latest version. The Palm site is constantly updated and improved, and the latest device ROMs will often only run on the latest versions of the Emulator.

Final Thoughts

Palm has been around for quite some time and they are still in their infancy. I'm sure that if you really look around your office, or pay attention to those next to you on the train or bus, you'll notice that many people use one Palm product or another. Palm devices have blazed a trail in the PDA world, and yet they are still producing newer and more advanced units every year. With all the applications built specifically for these devices, you can't go wrong with a Palm unit.

CHAPTER 6

Pocket PC

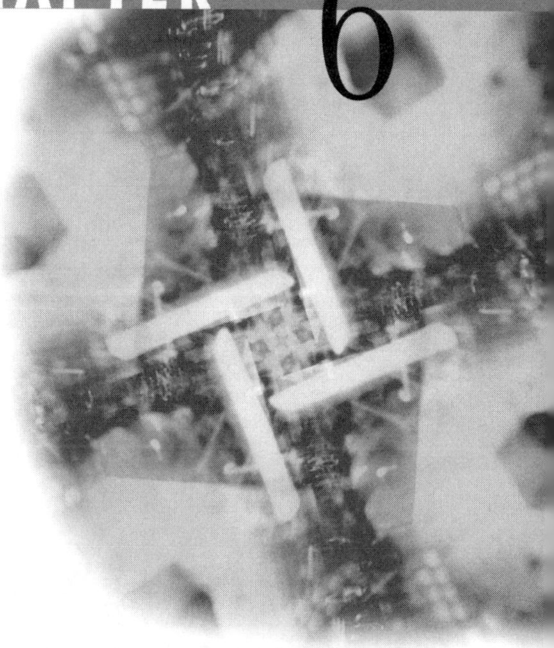

The second most common wireless devices, in our opinion, are Microsoft's Pocket PCs, which are Windows CE based. Palm currently has more market share, but Pocket PC-based handhelds are really gaining. We have a hunch that these devices will soon become more popular than Palm-based devices—only time will tell.

This chapter is dedicated to Pocket PC devices. The main Pocket PC Web site is, of course, from Microsoft and can be found at www.pocketpc.com. This site is packed with Pocket PC information with links to many other interesting areas.

Overview

The Windows CE operating system was designed as a platform for creating mobile computing devices in different shapes, sizes, and degrees of ruggedness. WinCE manages the interaction between application software and the hardware on the physical units. Figure 6.1 gives a good view of the many different WinCE devices available. They range from Pocket PCs, Handheld PCs, special phones, and rugged, custom devices.

Microsoft provides the software platform: Windows CE operating system, applications that integrate with Office, Outlook, Exchange, SQL Server, and others, while business partners provide the devices, peripherals, wireless networks, applications, and so on. Microsoft's mobility strategy is based on its

Figure 6.1 WinCE-based devices.

many business relationships, thus allowing a partnership in providing overall end-to-end business solutions.

Microsoft's original vision from conception in 1975 was to put a personal computer on every desk and in every home. In 1999, their vision changed to empowering people through great software any time, any place, and on any device. Microsoft might have said it, but the entire computing industry took notice and has been affected. We believe the wireless momentum has begun (hence, this book) and will overpower the computing industry in just a few years. Before we get there, let's look at Pocket PCs and some of their features.

Pocket PCs have not been around for as long as Palm devices. They have only been around for a handful of years, but are gaining amazing momentum in the marketplace because they are small versions of desktop computers. Palm devices were originally based on personal organizers looking very similar to several handheld PCs, but far from having their capabilities.

The applications look and feel very similar to Windows 95 or 98, with the Start menu on the taskbar and drop-down lists just like desktops. The usage transi-

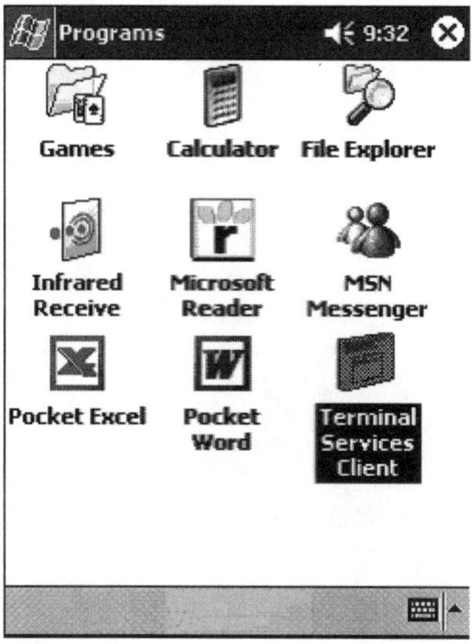

Figure 6.2 Typical Pocket PC screen.

tion to these units is fairly easy because of this similarity. Figure 6.2 shows a typical screen shot with the Programs listed in the taskbar. To close a window, simply tap on the X, top right, just as you would with Windows on the desktop.

Physical Units

The units discussed in Table 6.1 are Pocket PC based and are some of the most popular units available today. Unlike Palm, who manufactures their own units, these units are not built by Microsoft but by OEM partners. Just as Microsoft doesn't manufacture personal computers or servers, Microsoft does not manufacture the physical Pocket PC units either.

Device Comparison

Table 6.1 gives a great comparison between the physical units mentioned previously. Review the model differences and key points to figure out which one would best suit your needs. Our choice for all the examples throughout this book is the Compaq iPAQ H3870 unit.

Table 6.1 Microsoft's Devices Comparison Chart

MODEL	MEMORY	PROCESSOR	DISPLAY	BUILT-IN EXPANSION	SYNC OPTIONS	BATTERY LIFE
Audiovox Maestro PDA-1032	ROM 32MB RAM 32MB	206 MHz, Intel StrongARM 32-bit processor	Type: Reflective TFT LCD Number of Colors: 65,536 Resolution: 240 × 320	CF II and SD card slots Add-on expansions: third party	USB cradle, infrared, wireless phone connector cable, AC adapter, soft case, stylus (3), CD-ROM	Up to 8 hours
Compaq iPAQ Pocket PC H3760/H3765	ROM 32MB RAM 64MB	206 MHz, Intel StrongARM 32-bit processor	Type: Reflective TFT LCD Number of Colors: 4,096 Resolution: 240 × 320		USB cradle, infrared, wireless phone connector cable, AC adapter, soft case, stylus (3), CD-ROM	Up to 9 hours
NEC MobilePro P300	32MB flash ROM 32MB DRAM, 32MB SD Card included	206 MHz, Intel StrongARM 32-bit processor	Type: Reflective 3.8" QVGA TFT LCD Number of Colors: 65,536 Resolution: 240 × 320	CF type II, SD, Slave(function)	USB cradle, infrared	Up to 10 hours (estimated)
Compaq iPAQ Pocket PC H3870/H3835	ROM 32MB RAM 64MB	206 MHz, Intel StrongARM 32-bit processor	Type: Reflective TFT LCD Number of Colors: 65,536 Resolution: 240 × 320	SD card slot for memory expansion	USB cradle, infrared, wireless phone connector cable, AC adapter, soft case, stylus (3), CD-ROM	Up to 12 hours
Toshiba e570	ROM 32MB RAM 64MB	206 MHz, Intel StrongARM 32-bit processor	Type: A-Si reflective TFT LCD Number of Colors: 65,536 Resolution: 240 × 320	CF Type II and SD	USB cradle, infrared	Up to 8 hours

Table 6.1 (Continued)

MODEL	MEMORY	PROCESSOR	DISPLAY	BUILT-IN EXPANSION	SYNC OPTIONS	BATTERY LIFE
Casio E-200	ROM 32MB RAM 64MB	206 MHz, Intel StrongARM processor	Type: Reflective TFT LCD Number of Colors: 65,536 Resolution: 240 × 320	CF Type II/SD/ infrared/USB/ serial port	USB host available in cradle, PC Card Sled, and adapter (cradle ships with unit, PC Card Sled and adapter are optional), infrared	Up to 10 hours
HP Jornada 560 Series	ROM 32MB RAM 32/64MB	206 MHz, Intel StrongARM 32-bit processor	Type: Reflective TFT LCD Number of Colors: 65,536 Resolution: 240 × 320	CF Type 1 extended	USB cradle, infrared	Up to 14 hours
Fujitsu-Siemens Pocket LOOX	ROM 32MB RAM 64MB	400 MHz	Type: Reflective TFT LCD Number of Colors: 65,000 Resolution: 240 × 320	Integrated SD/ MMCard Slot, CF Card Slot Type II	USB/Serial cradle, infrared	Up to 12 hours
mm02 XDA	ROM 32MB RAM 32MB	206 MHz, Intel StrongARM 32-bit processor	Type: Reflective TFT Number of Colors: 4,096 Resolution: 240 × 320	CF Type 1 extended	USB cradle, infrared	Up to 14 hours
Legend XP100	ROM 32MB RAM 64MB	206 MHz, Intel StrongARM 32-bit processor	Type: Reflective TFT Number of Colors: 65,536 Resolution: 320 × 240	CF Type 1 extended	USB cradle, infrared	Up to 10 hours

Courtesy of Microsoft

Special Units

There are also durable Pocket PCs for nearly all weather purposes.

The Symbol PPT 2800 can tolerate temperatures from –4 to 122 degrees Fahrenheit (–20 to 50 degrees Celsius). It is rugged enough to withstand multiple FOUR-foot drops, is sealed against rain and dust, has an integrated laser bar and code scanner, and integrated WLAN/WAN radio options.

Units in the Intermec 700 Series have integrated radio and scanning options, a rugged design for harsh environments, a numeric keypad, application auto-setup and configuration, and remote device management.

The Symbol PDT 8100 can tolerate temperatures from –4 to 122 degrees Fahrenheit (–20 to 50 degrees Celsius). It is rugged enough to withstand multiple 4 foot drops, is sealed against rain and dust, has an integrated laser bar and code scanner, and integrated WLAN/WAN radio options.

Compaq iPAQ H3870 Unit Specs

Let's have a closer look at this unit. This is a very new device, so new that when we tried to obtain one, it had to be ordered directly from the manufacturer, as none were available in the stores. Figure 6.3, taken from the Compaq CD that comes with the unit, shows the unit's basic components.

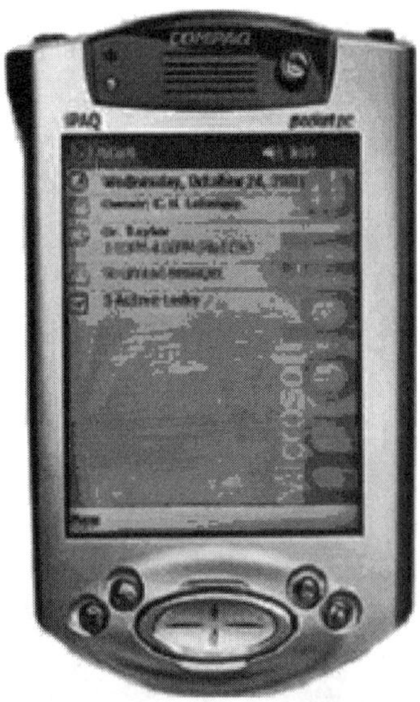

Figure 6.3 Compaq iPAQ H3800 series.

The Compaq iPaq has a stylus just as Palm devices do, but its location and design are so advanced that the convenience is delightful.

Then we have the power indicator, which changes color based on whether the unit is charged, charging, or if an alarm is set—very visual. Next to the power indicator is the Power button. You might notice at times that when you push it to shut off the unit, the screen still shows something but without the backlight. We find that this happens when something is still running. Simply close the application, push the button again, and the unit will shut off properly.

Directly under the COMPAQ label is the speaker, which is quite nice and will be used since the units have a media player, which allows music and video. Then we have the color display screen, which is large, and has a backlight that really displays the screen beautifully. The screen has 65,536 colors with 240 × 320 resolution. Remember that the backlight will drain the battery fairly fast, so be sure to set the backlight-to-turn-off indicator to a low number. The default is 30 seconds.

Next are the four programmable application buttons. As seen in Figure 6.3, from left to right they are Calendar, Contacts, Inbox, and iTask. iTask is very useful because it shows all currently active applications. One problem we find with this unit is that even after you close an application via the x-out button, the application still remains active. To deactivate it, simply press the iTask, tap and hold on any item that appears in the pop-up until a dotted circle appears, and then choose the *close all tasks* option—do this by tapping it once. All executing applications will end, freeing up memory. If too many applications are running at once, the units slow down considerably.

The big oval button on the unit is the Navigation button. Press it to scroll up, down, left, or right.

The last button is on the unit's left side. The out-of-the-box default for this button is to run the recording application. Press it and after the beep, you can record conversations or anything else you desire. This button can be reprogrammed to do whatever you want.

Directly above the word *iPAQ* on the front of the unit is a small sensor that controls the backlight intensity. It senses your lighting environment and adjusts the backlight as required (nice!!).

On the top of the unit is the stereo headphone jack, and next to that (more on the front than on the top of the unit) is another indicator light used for Bluetooth, if the unit has this capability. When Bluetooth is active, the indicator will flash blue.

The last feature is the small, barely visible microphone directly above the light sensor. It is small but quite sensitive and picks up sound with great detail. The top of the unit has more features. From the top you can see the headphone jack (as mentioned earlier), another microphone, a hidden infrared port for beaming information to other devices, the stylus, and finally the Secure Digital (SD) expansion slot. This slot holds a small, stamp-sized expansion card. We're using a 64MB multimedia memory card from SanDisk. That's 64MB of extra storage space on a tiny little card—amazing!

The bottom of the device also has some interesting features. First is the charging and communications port. It connects to the cradle or cable for connectivity to the desktop. Then there's the small Reset switch. Unlike Palm, which makes it difficult to press the Reset switch by accident, the Compaq iPAQ's switch is easily accessible, which is good and bad. Press the Reset switch to simply restart the unit if it freezes, which it does quite often.

A hard reset feature is also available. Performing a hard reset will flush files from the unit's memory. To perform the hard reset, press and hold the two outside application buttons, insert the stylus into the Reset switch area, and press the switch for five seconds. Then, to reactivate your unit, insert the stylus into the Reset switch again and press it for one second or simply connect the unit to the cradle. If you ever forget your device password, you'll have to perform a hard reset.

Last is the expansion pack connector. This large area allows for the unit to connect to an expansion pack for PC Cards and CompactFlash Cards.

The cradle for the H3870 is used to synchronize to the desktop and to recharge the unit. It comes with a USB and serial connection cable. Once the device is put into the unit, Microsoft ActiveSync on the desktop immediately synchronizes the units based on the synchronization parameter settings, as we'll see shortly. Of course, the desktop must be powered on.

Software

The following is a list from Microsoft of all the features and applications available on Pocket PCs:

Desktop Synchronization. Microsoft ActiveSync technology supports continuous or on-demand synchronization.

Infrared Send and Receive. Support for beaming contacts, schedule, or any file including images, music files, or documents.

Network Connections. Automatic management of work or home connections.

Email. Microsoft Pocket Outlook provides full support for POP3 or IMAP4 email, including HTML messages, meeting requests, Word and Excel attachments, plus voice messages.

Instant Messages. MSN Messenger includes automatic logon, multiple chat sessions, online buddies, privacy controls, and My Text quick replies.

Online. The multimode Pocket Internet Explorer supports any HTML or WAP and secure sites as well as XML Web services.

Offline. Offline browsing available with Pocket Internet Explorer or with integrated AvantGo.

Macromedia Flash Support. Available as a free download from Macromedia.

Handwriting. Rich Ink and Transcriber, a natural handwriting recognition application.

Character Recognizing Input. Letter and block printing recognizer for inputting characters naturally or in an input method familiar to users of Graffiti.

Integrated Voice Recorder. All Pocket PCs include a voice recorder for recording spoken notes or even responding to email.

Configurable Today Screen. A fully configurable Today screen that allows users to view their most critical data. Personalized background images and colors can also be configured.

Notifications. A powerful notification engine lets users know about new email or instant messages, appointments, and critical events through displays, sounds, or integrated hardware lights.

Device Security. Two levels of power-on password protection, including strong password support for enhanced security.

VPN Client (PPTP). Integrated support for virtual private network connections.

Terminal Services Client. A full mobile version of Terminal Services Client provides for remote access to Microsoft Windows-based servers to run applications or perform server maintenance.

Multimedia. Windows Media™ Player supports audio and video playback, including MP3 files. Also provides for playing media on the device, or streaming content from the Internet or intranet. You can play music in the background while working within other apps.

eBooks. Microsoft Reader with ClearType technology delivers clear, crisp, highly readable text. Includes an integrated Audible player. Digital rights management support for the latest and greatest titles.

Gaming. Solitaire is included on the device, and many third-party high-resolution, action games with arcade-style controls are also available.

Microsoft Word. Pocket Word is a robust mobile version of Microsoft Word supporting rich formatting and graphics, plus other features including spell check and word count.

Microsoft Excel. Pocket Excel is a robust mobile version of Microsoft Excel supporting virtually all standard features.

Desktop Personal Information Manager. A full desktop version of Outlook® 2002 ships with every Pocket PC 2002.

Desktop Software. Every Pocket PC ships with a CD that includes full desktop versions of ActiveSync, Internet Explorer, Reader, and Windows Media Player (as well as many other Pocket PC products) at no extra charge.

As you can see, Pocket PCs have many features. These units are not personal organizers but full-blown mini-personal computers. We're constantly amazed at all the functionality and potential these units have to offer.

Microsoft ActiveSync

Now that we're familiar with the units, let's get our H3870 unit connected to the desktop. This will grant the device the capability to synchronize schedules, calendars, notes, email, applications, and the like between the desktop personal computer and the wireless device. This is done via Microsoft ActiveSync. Our units were shipped with ActiveSync version 3.5.

To connect to the desktop and synchronize files, you'll have to run the Compaq iPAQ Pocket PC Companion CD that comes with the device on your desktop. The CD's main executable will automatically begin and display the introduction screen. Click the Play button, and on the next screen, click on the *Start Here* area to bring you to the Start Here window.

Before you begin, ensure that your cradle is not connected to the desktop. You should not connect the two until after Microsoft ActiveSync is installed.

The installation application suggests upgrading Microsoft Outlook to Outlook 2002. We have a laptop with Outlook 2000 installed and didn't want to alter our Outlook settings by performing the upgrade. Therefore, we decided to bypass the Outlook upgrade even though the application suggests not to. We've found no problems with the synchronization process so far, so we can report that not upgrading Outlook if you have the 2000 version will not be a problem. Mind you, if you do not have Microsoft Outlook on your desktop, you must install the version on the CD. ActiveSync needs Outlook to function and therefore both must be present on your machine.

Read the Overview window and proceed to installing ActiveSync 3.5.

NOTE

Microsoft's Pocket PC synchronization application is called Microsoft ActiveSync, while Palm's synchronization application is called HotSync.

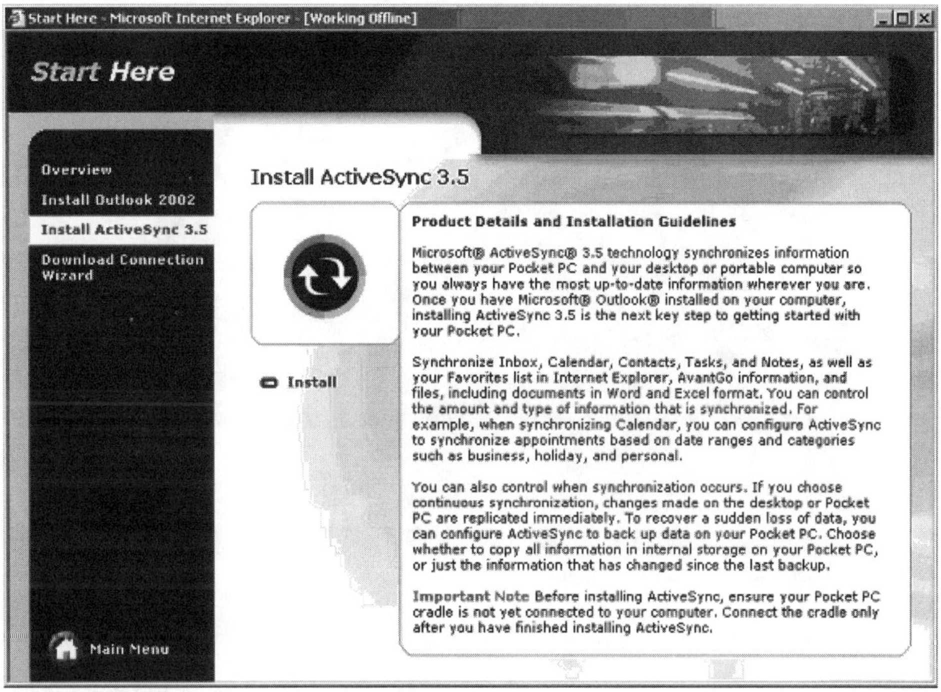

Figure 6.4 Start Here—Install ActiveSync 3.5.

Figure 6.4 shows the startup window for the installation of ActiveSync. The process begins by clicking on the install link in this window. You should choose to run the program from its current location and confirm that you want to install and run Microsoft ActiveSync 3.5. Installation will begin after that.

Once you have done that, you will need to determine where the ActiveSync directory will be located. Choose the default as shown in Figure 6.5, and click Next to begin installation.

Microsoft ActiveSync 3.5 is now installed and the device's cradle can now be connected to the desktop to begin the synchronization process.

The Get Connected window, shown in Figure 6.6, will open after installation is complete. If you do not connect the cradle and click the Next button, the system will attempt to find the device, not be able to, and return several error screens. Don't worry—simply connect the cradle and the software will recognize that you've done so, register a connection, and continue on its merry way.

The next step is to create a partnership between the desktop and the mobile device (see Figure 6.7). This will allow the synchronization to happen

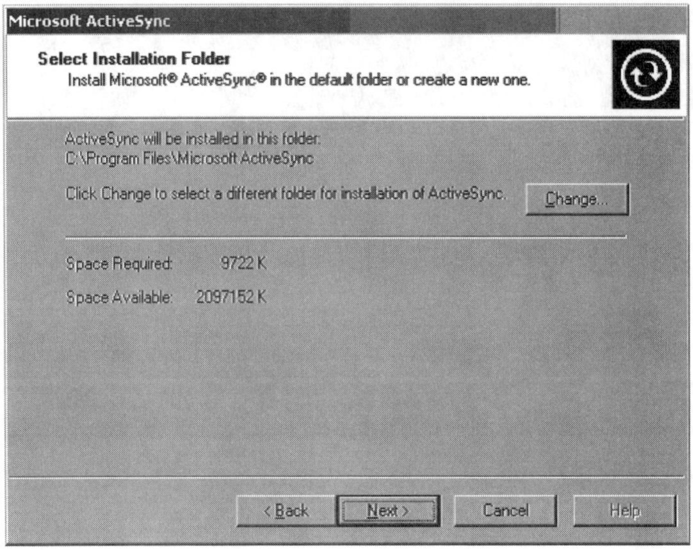

Figure 6.5 Choose Installation Folder.

between each. First, you must select what you'd like to synchronize as shown in Figure 6.8.

One of the problems we had in setting up this option was that our Microsoft Outlook is set up to function from our office network. This means we have no

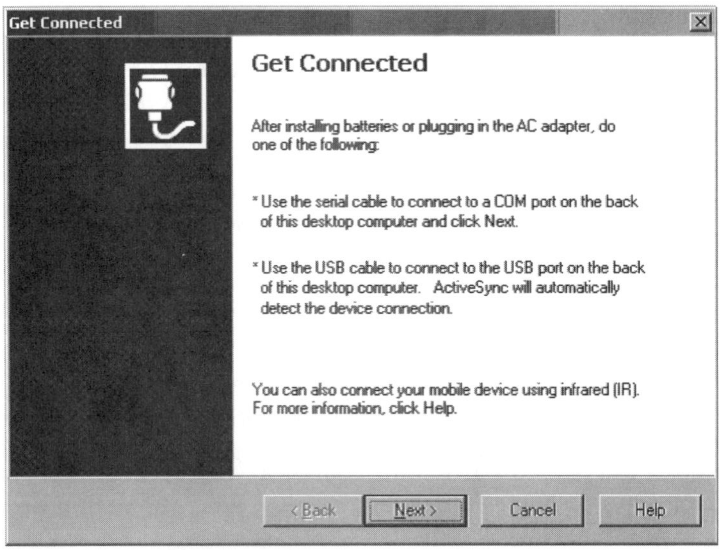

Figure 6.6 Get Connected.

Figure 6.7 Setting up a new partnership.

entries in the Calendar, Contacts, Inbox, or anywhere if we're not connected to the network. Therefore, to avoid errors in setup and connectivity, we simply unchecked all boxes except for the Favorites and continued on.

Click the Next button and the setup is complete. With the mobile device connected through the cradle to the desktop, and Microsoft ActiveSync installed,

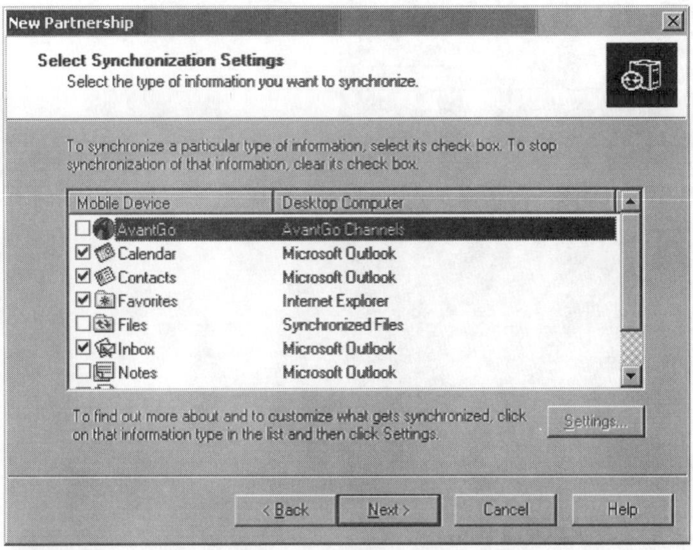

Figure 6.8 Selecting synchronization settings.

Figure 6.9 Microsoft ActiveSync icon.

synchronization will automatically take place. A nice feature is that each time the unit is placed in the cradle, synchronization will automatically take place. There's no need to manually execute applications—it's done immediately. However, if you'd like to run ActiveSync manually, simply click on the Taskbar icon as shown in Figure 6.9 and choose the synchronization option.

We've found that sometimes the desktop does not, for some reason, recognize the mobile device after it has been placed in the cradle, even though all seems properly connected and defined. Therefore, we tried changing the port from the default COM3 to COM4 in the Connection Settings area, as shown in Figure 6.10, and all worked fine from then on. However, this only helps if you're using the Serial connection.

A very nice feature about using a Microsoft-based mobile device is that it's fully integrated with the desktop. From Explorer on the desktop you can view

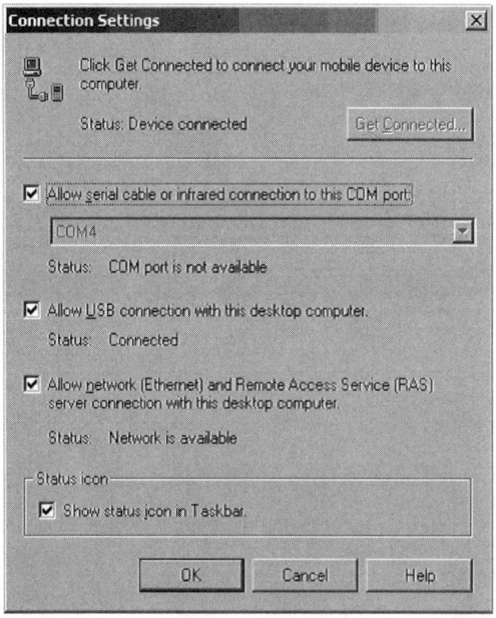

Figure 6.10 Connection Settings.

all files on the mobile device (see Figure 6.11). This is very handy, because viewing files on the mobile device doesn't allow you to see file extensions. Look for the *Mobile Device* entry in Explorer and treat it as any other drive on your system. Another nice feature is that all files on your storage (memory) card, if you have one, can also be viewed.

If ever you want to copy a file from the desktop to the mobile device, simply copy it via Explorer over to the desired directory on the mobile device. You will be doing this if you're building applications for Pocket PC. Many tools copy all components on the first install, but fail to correctly handle future upgrades or changes. In that case, you'll have to manually copy files to the device.

Now we have all the software installed to connect the desktop and mobile device. We know how to synchronize our Inbox, Outlook applications, and lots of other information. The next step is to set up specific synchronization parameters.

Get into the Microsoft ActiveSync application and click on the Options icon. Then, choose whichever application you want to synchronize by highlighting it with the cursor. For our example in Figure 6.12, we've chosen the Calendar option.

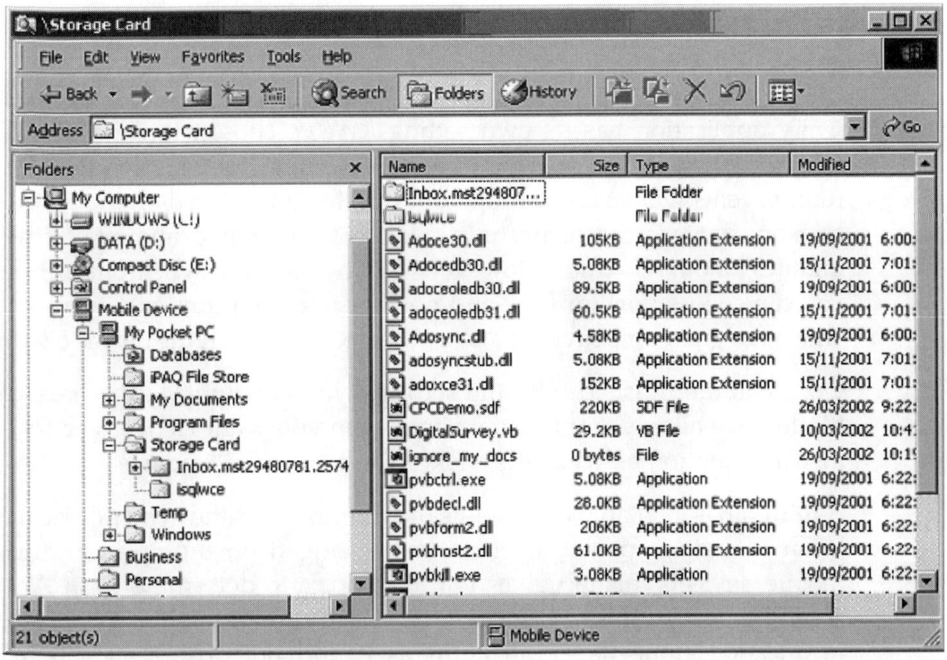

Figure 6.11 Explorer and mobile device.

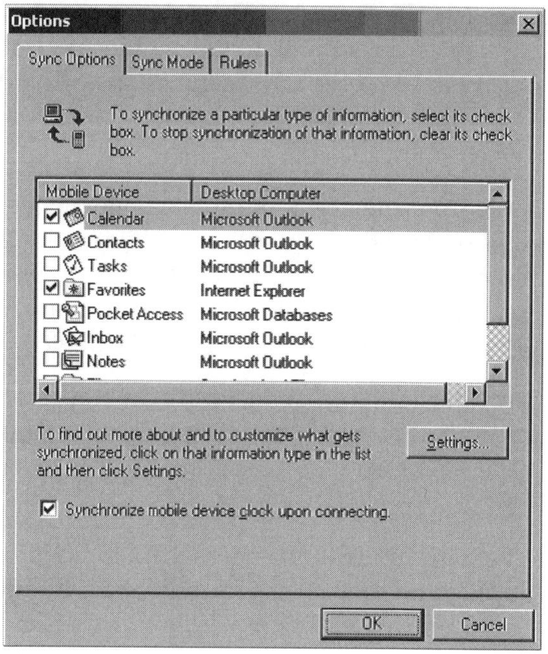

Figure 6.12 Synchronization options.

Click the Settings button to jump to the settings for Calendar (see Figure 6.13). Every application has its own settings. We've chosen to show the Calendar option because if you have lots of entries in your desktop calendar, all past and future events will be taken from your desktop and placed on the mobile device each time you synchronize. This can take a considerable amount of time (5 to 10 minutes for our schedule). However, with the Calendar synchronization settings, you can choose how far back and how far into the future you want to upload, which is a great time saver. It doesn't waste memory on the device either.

Once you are finished with the settings, you can jump to the next tab, Sync Mode. This allows you to determine when you want ActiveSync to synchronize information between the desktop and the mobile device.

A third tab is available that allows the user to select the synchronization rules. What would happen if an item were changed on the desktop and on the mobile device? This tab gives you the choice to determine what ActiveSync should do under this circumstance. You can choose to either leave the entries in question alone (leave all items as is and not do any synchronization),

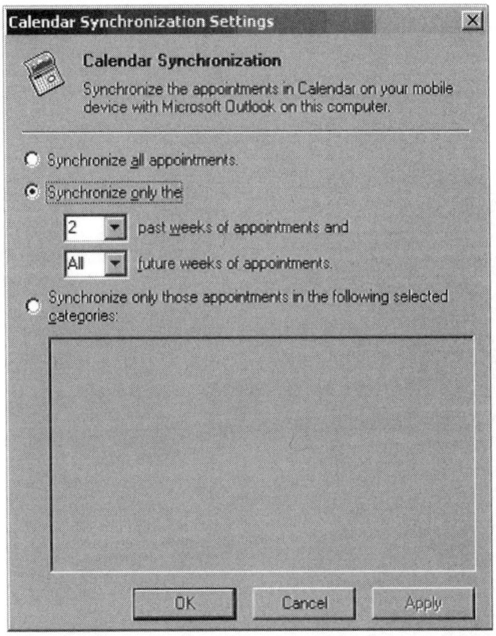

Figure 6.13 Calendar synchronization settings.

replace the item on the mobile device with the item on the desktop, or replace the item on the desktop with the item from the mobile device. Quite an important feature, wouldn't you say?

Another great feature of Microsoft ActiveSync is the built-in capability to invoke backups and to restore the information. Click on the Tools icon once in ActiveSync and choose Backup/Restore. You'll see all the backup options available to you, as shown in Figure 6.14.

That's about all there is to Microsoft ActiveSync 3.5. A very simple installation to quickly and easily set up and run synchronization between the desktop and the mobile device. Now you can safely remove the device from its cradle and be wirelessly mobile.

For more information on the H3870 device, use Explorer to view the CD contents. You'll find a Documents folder with a lot of the information you might be looking for.

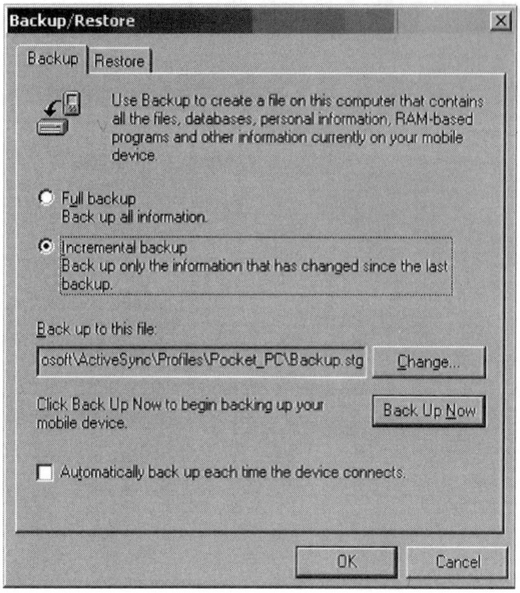

Figure 6.14 Backup/Restore options.

Beaming

Let's now quickly review beaming information between your mobile device and the Desktop, or another mobile device.

The Compaq iPAQ H3870 unit we're focusing on in this book has an IR port. This allows the unit to send and receive information via this port with other units or the Desktop. Simply point the unit's IR ports toward each other and they should register a connection—it's as simple as that, right out of the box. When we placed our Compaq iPAQ H3870 unit next to our Toshiba ThinkPad T22 just as we had done with the Palm m500, the same thing happened—we heard a beep, and a screen came on asking if we wanted to transfer a file. We tapped on a file and over it went.

On the Compaq unit, to beam an application or file, simply tap the Start button, select File Explorer (if you put it in the main menu selection), tap on it, and then select a file. Tap and hold the stylus on a file and a pop-up menu will appear. Choose the *Beam File* option and you're done. While doing this, you must, of course, point the unit to the receiving device and wait until the Status box confirms the transfer. To receive, point the unit to the sending device and follow the same simple instructions.

Final Thoughts

Pocket PCs are mini handheld computers. They look and feel just like small versions of Windows 95, 98, or 2000. These are great devices with lots of potential. We really enjoyed using them and building applications for them.

The Compaq iPaq H3870 series is an amazing machine. All of our Palm porting coworkers were quite wide-eyed about this device and so were we. We just all wished that the units were priced lower so they'd be more affordable.

CHAPTER 7

Mobile Application Development Tools

Currently in the marketplace, two operating systems are powering most of the portable devices: Microsoft's Pocket PC and Palm OS. (There are attempts to bring Linux and Embedded Windows XP into this arena, but it will take a long time before any of the portable OS newcomers take one of the leading places in the OS market.) Some of the tools that we will talk about in this chapter are capable of producing code for both of these major mobile device operating systems. We will create a small application with each of the tools and use it to highlight some differences, advantages, and disadvantages for each tool described. So, let's start with Microsoft's tool, called eMbedded Visual Basic 3.0.

Microsoft eMbedded Visual Basic

If your target platform is Windows CE (Pocket PC) based, then eMbedded Visual Basic (eVB) is one of the choices that you should consider. If you are familiar with Visual Basic you will especially enjoy building portable applications with this tool. You can download Microsoft embedded Visual Tools 3.0 (this is the official name for the product) from www.microsoft.com/mobile/developer/downloads/emvt30/. Please keep in mind that this download is more than 300MB in size and that download could take many hours to complete.

System requirements for the installation of this tool are as follows:

- PC with Pentium 150 MHz or higher processor recommended.
- Microsoft Windows 98 Second Edition, Microsoft Windows NT Workstation operating system version 4.0 with Service Pack 5 or later, or Microsoft Windows 2000 operating system.
- Recommended 48MB of RAM.
- Hard-disk space required: Minimum installation: 360MB; complete installation: 720MB.

eVB developers can create applications with the code that is a subset of the Visual Basic language. Many language features that you can find in Visual Basic 6.0 are not supported in eVB 3.0. For example:

- All variables are variant by default.
- eVB can only host ActiveX controls.
- Several Visual Basic controls are not supported and some properties are missing.

Once you compile your application, the code produced is a P-code version, which requires the device to support its own runtime in order to be interpreted. As you will see, the only supported platforms are Windows CE based.

Once you have selected the type of device you will be developing applications for, you will be presented with an Integrated Development Environment (IDE) similar to the one shown in Figure 7.1. This screen resembles Visual Basic IDE in many ways, but the moment you open the default form, it becomes clear that this product is targeted at mobile devices. For example, default size and format of the form is such that it represents a WYSIWYG experience for the developer.

We can now begin to create the famous "Hello World" application with eVB. The only tasks we will perform are changing the form caption message, adding one label, and changing the label caption message to "Hello World".

Once you have completed this you can deploy the application from your Tools menu. If your mobile device is connected and ActiveSync is installed, you should see a picture similar to the one in Figure 7.2.

One of the nice features of eVB is that it allows developers to capture the mobile device screen (not only eVB-related screens) via the Tools->Zoom application that is provided with it.

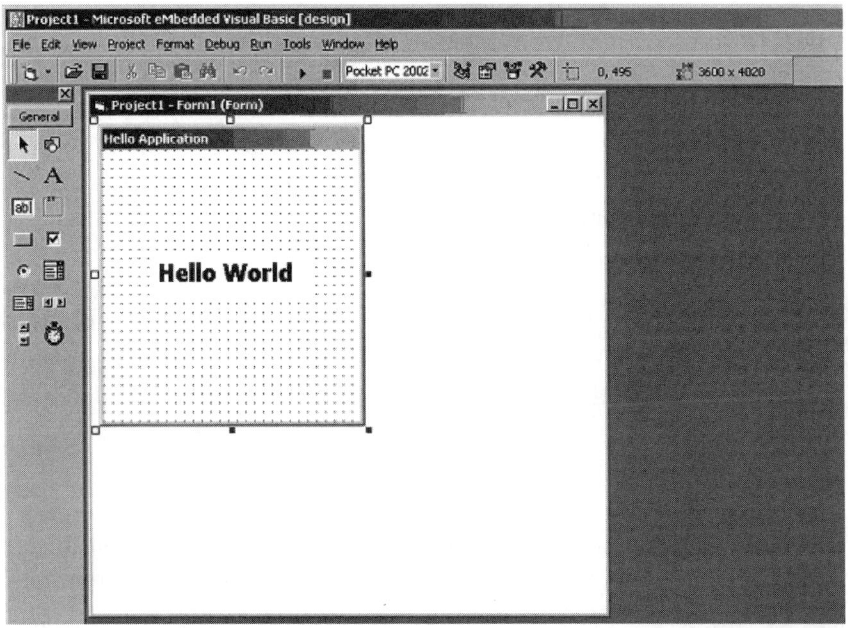

Figure 7.1 eMbedded Visual Basic 3.0 "Hello World".

Developing for Windows CE powered devices with eVB is extremely fast. In addition, don't forget that you can connect from an eVB database to an ODBC-compliant database (e.g., DB2 or Sybase ASA) and natively to Microsoft SQL 2000 CE edition. Microsoft does have two more tools available in the "arsenal" for building mobile solutions: eMbedded Visual C++ and Visual Studio .NET.

Figure 7.2 eMbedded Visual Basic 3.0 device screenshot.

AppForge 2.1.1

AppForge 2.1.1 is another development tool based on Microsoft's core product. This mobile development platform supports both players in the mobile arena. You can download an evaluation copy from www.appforge.com. Since AppForge is an extension of Visual Basic 6, minimum requirements for this development toolkit are similar to the Visual Basic requirements:

- Windows 95, 98, NT, 2000, ME, or XP computer (95/98/ME and NT must have Windows Installer installed).
- 32MB RAM, Pentium 90 MHz processor, and 40MB of available hard drive space.
- Microsoft Visual Basic 6.0 (Service Pack 5).

Once installed, this tool becomes part of the Microsoft Visual Basic 6.0 IDE. AppForge 2.1.1 will provide you with the set of ActiveX controls (or *Ingots* in AppForge terminology). AppForge will also allow you to manage database files, fonts, graphics, and movie files. It provides the universal conduit tool that will allow exchange of data between portable devices and any ODBC-compliant database. One of the advantages of AppForge is that you do not need the emulation tool to deploy your application to the portable device itself in order to test it. To run AppForge built applications, you must install AppForge Booster, an operating system extension that enables AppForge written applications to run on the target devices. Booster extensions must be installed before you attempt to run any AppForge application on your target device. For the Palm-powered devices, you will need HotSync installed, and for the Pocket PC-based devices, you will need Microsoft's ActiveSync.

Since the process of developing the cross-platform application with AppForge is really just a couple of clicks away, you can have applications for Palm and Windows CE-based devices up and running very quickly. Again, it's time for our "Hello World" Application. Please note that in Figure 7.3 the signs for the Microsoft Visual Basic label control (left-hand side of the general tool bar) and the AppForge label toolset control (right-hand side of the general tool bar) have been circled.

You can only use the AppForge installed objects; otherwise, you will receive a VB error that will warn you that the object is not supported in AppForge projects. To make sure that your application works, you should perform two tests. The first test is to simply run the application. The resulting screen is pictured in Figure 7.4. For most of the applications that do not use specific API calls (system, database, or otherwise), this test will tell you immediately whether your application will work.

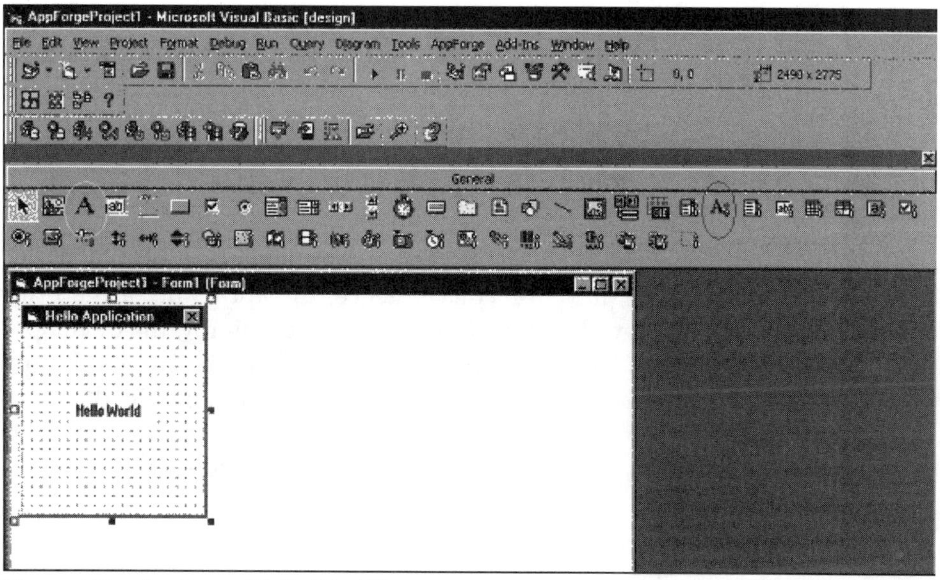

Figure 7.3 AppForge "Hello World".

The second test is to deploy the application to the mobile device of your choice. We assumed here that an applicable (Palm or Windows CE) version of Booster application is installed already. Once that is completed, the "Hello World" application can be deployed (compiled and installed) to the Palm device.

In our example, we will deploy the application to the Palm Emulator Program to capture a screenshot of the application. It should look very similar to Figure 7.4.

AppForge is an excellent choice for mobile application development, especially if your in-house talent (developers) have Visual Basic 6 experience. The learning curve required is minimal and developers can start productive design within weeks. It is important to note that AppForge uses Palm PDB database format to store data on the mobile device. AppForge also offers one of the best

Figure 7.4 AppForge "Hello World" executed locally.

conduits for the data exchange between mobile device and desktop or enterprise class database. You can even specify bi- or single-direction synchronization and selectively synchronize tables.

IBM Everyplace Mobile Application Builder 7.2.1

IBM's relational database engine has gone full circle. From the mainframe to the mobile device, it is everyplace; hence, the apt name of the product that runs on mobile devices—DB2 Everyplace. Accordingly, the development tool that supports this environment is called DB2 Everyplace Mobile Application Builder. This toolkit is primarily designed with Palm mobile devices and ease of DB2 database integration in mind. To run Mobile Application Builder, your workstation must satisfy the following hardware and software requirements:

- A Pentium II processor or higher.
- At least 128MB of RAM.
- 20MB of disk space for installation of Mobile Application Builder.
- 75MB of disk space for installation of additional tools required by Mobile Application Builder.
- Windows NT 4.0, Windows 2000, or Window XP Professional.
- Development Software: Cygwin; GNU PRC-Tools 2.0, Palm SDK 4.X; PilRC.
- Palm OS version 3.0 powered device or higher.

You can download DB2 Everyplace Mobile Application Builder and DB2 Everyplace database engine at www14.software.ibm.com/webapp/download/search.jsp?go=y&rs=gnutools.

You will also need to download additional required software (to compile your code and create a re-distributable Palm packages). Following are the links to the download sites:

http://sources.redhat.com/cygwin/ for **cygwin-b20.1-full.exe**

http://sourceforge.net/projects/prc-tools/ for **prc-tools2.0.exe**

http://sourceforge.net/projects/pilrc/ for **pilrc.zip**

www.palmos.com/dev/tools/sdk/sdk2452534/sdk40.html for **Palm OS SDK 4.0**

At this point, you should have all of the required components installed and have read the ReadMe files. We are now ready to start our Mobile Application Builder. The first screen you will see, once you select New Project from the File

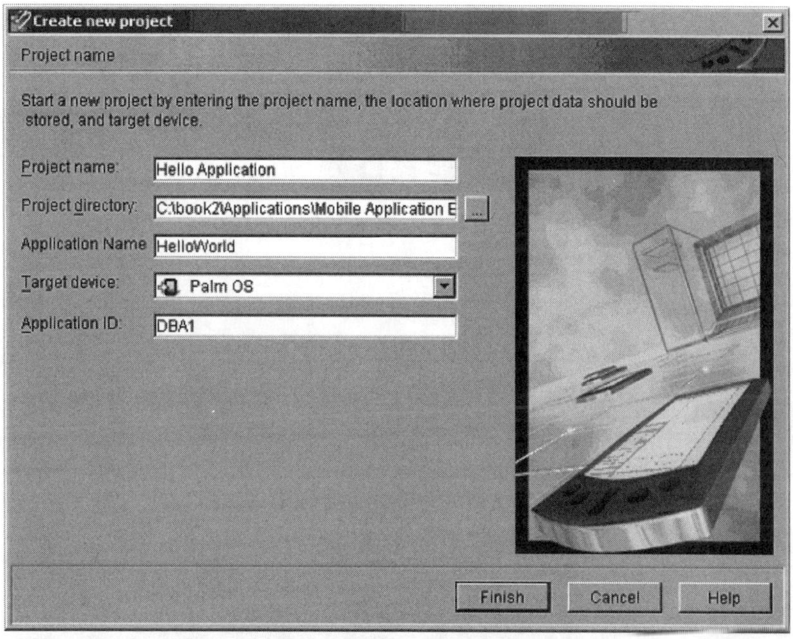

Figure 7.5 Mobile Application Builder project start.

menu, is similar to the one shown in Figure 7.5. As you can see, the only supported Mobile Operating System at the moment is Palm OS. Once you complete all of the fields and you click the Finish button, you will see a picture similar to Figure 7.6. Again, minimum effort is required to change the form caption, insert one label, and change the label caption to "Hello World". It's an easy task since the toolboxes are similar across all of the tools.

Once we compile the application and deploy it to the Palm-powered mobile device, you will see a screen on your mobile device similar to the one in Figure 7.4. DB2 Everyplace Mobile Application Builder can also help you improve the development cycle of DB2-related applications. Mobile Application Builder will use Data Management Language (DML) to create appropriate table definitions (in our example we can use DML definition for the TESTERS table in Listing 7.1). Once the table definition is accepted (even comments in the DML statement will create an error), data aware forms are only couple of clicks away.

You will notice in Figure 7.7 that one new item has come up since we added the table definition. On the left-hand side of the user's IDE, you will see a new table called testers under the "Tables" item. From this point on, database functions such as Add, Delete, Update, and Insert for all of the fields can be created with only a few clicks. That is powerful, and if you want to use this tool it would be good idea to use the DB2 Everyplace database engine since this combination will ensure quick implementation and shorter development times.

Figure 7.6 Mobile Application Builder "Hello World" application.

```
CREATE TABLE "DB2ADMIN"."TESTERS"    (
                  "FIRST_NAME" VARCHAR(40) ,
                  "LAST_NAME" VARCHAR(40) ,
                  "STREET" VARCHAR(100) ,
                  "APARTMENT" VARCHAR(10) ,
                  "CITY" VARCHAR(50) ,
                  "STATE" VARCHAR(20) ,
                  "POSTALCODE" VARCHAR(10) ,
                  "COUNTRY" VARCHAR(50) ,
                  "INTERNET_ADDRESS" VARCHAR(50) ,
                  "SOCIAL_SECURITY_NO" CHAR(11) NOT NULL )
```

Listing 7.1 DML script to create testers table.

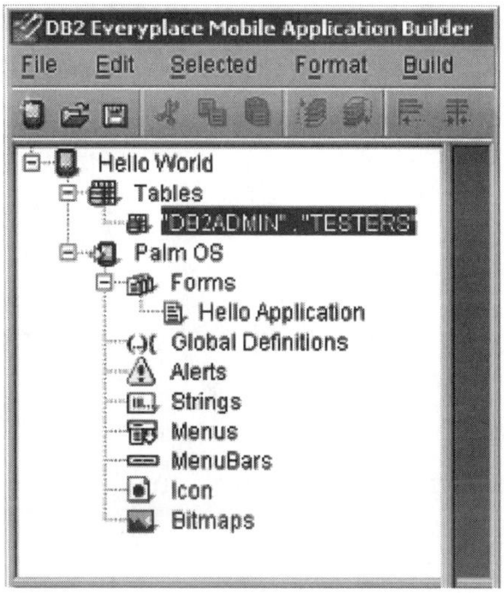

Figure 7.7 Mobile Application Builder database.

CASL IDE

Compact Application Solution Language (CASL) from Feras Information Technologies is one of the solutions that provides a complete RAD development environment. An evaluation copy of this excellent product can be downloaded from www.caslsoft.com/download.html.

One of the major parameters in evaluating products for mobile development is product coverage and code reusability. The last thing you as developer would want to do is develop and maintain separate versions of code for different mobile platforms. CASL IDE (or CASLide) allows you to create applications for Palm OS (2.0 and newer), Win32, and WinCE 2.11 from the same source. CASL scripting language is also simple to understand for developers who have previous experiences with Visual Basic or JavaScript. This means that the skills of your developers can be easily converted to cover mobile application development.

CASL is a mature product that has been around since 1996, and in order to run it you will need the following configuration for your development workstation:

- Any 32-bit Windows-based environment (Windows NT, Windows 2000, Windows XP, and Windows 98).
- Minimum Pentium II processor or higher.
- 128MB of RAM.

- About 40MB of disk space for the CASL and required additional components.
- HotSync for connection to your Palm-powered device, or ActiveSync for the connection to WinCE-based device.

In order to start development and deployment with CASL, you will also need following components:

- Cygnus package for Windows (download at www.caslsoft.com/gcc_tools/cygwin-b20.1-full.exe)
- PRC tools (download at www.caslsoft.com/gcc_tools/prc-tools-2.0.exe)
- Palm OS SDK (download at www.caslsoft.com/gcc_tools/palmos-1-2-3.1-sdks-1.zip)

With these components installed, we can start (all over again) to build the "Hello World" application. Once you start the CASLide application you will be presented with the screen shown in Figure 7.8. It is a straightforward, easy-

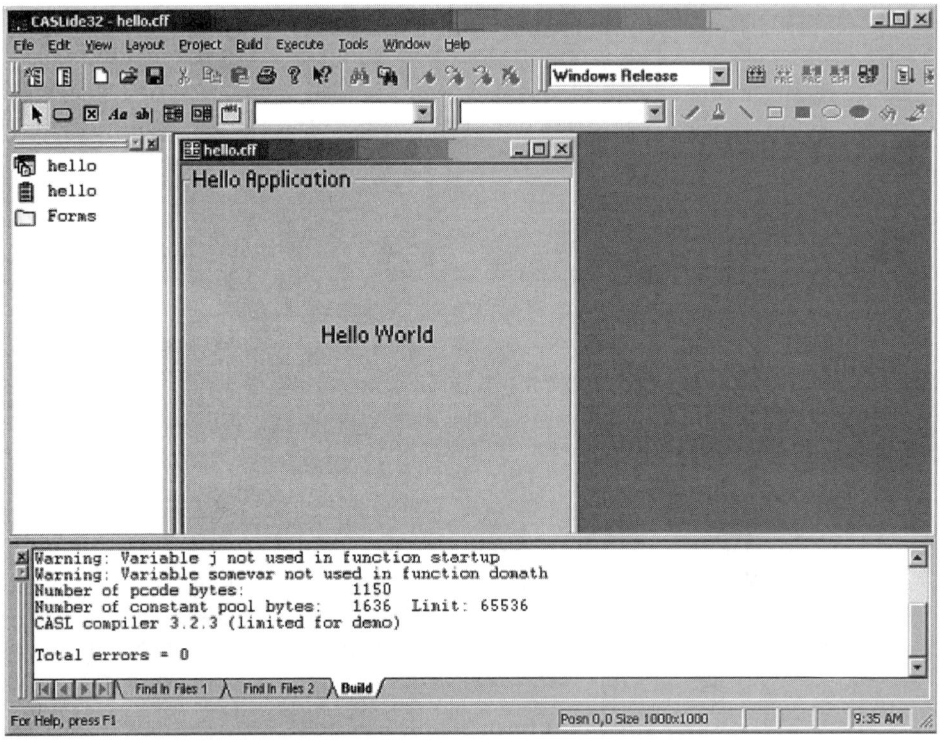

Figure 7.8 CASL "Hello World" application.

to-understand, and intuitive graphical user interface (GUI) that will enable you to be a productive mobile developer within two to three weeks. Once the application is up and running, just add the label object to the default form, change the caption to "Hello World", and change the frame caption to "Hello Application". From this point on, we can select the type of OS we would like to execute on. The application should look very similar to Figure 7.4 when run on the Win32 operating system or a Palm-powered device.

As previously stated, CASL is a "well rounded" product, and as such it offers a CASL conduit that allows you to use either ODBC (including enterprise class database engines such as Oracle, Microsoft and DB2) data sources or a CASL database (CASL database format is essentially CSV [comma separated] format with additional header information). Following is a short explanation of what can be achieved with a CASL conduit:

- Synchronize one-way or two-way with the PC during normal HotSync operation.
- Synchronize with any ODBC-compliant database on your PC.
- Synchronize with any CASL formatted database in your CASL HotSync folder.

The CASL conduit can be freely distributed. CASL also provides the following utilities that will help the developer distribute the applications successfully:

- CozndReg installs CASL conduit and registers with Windows OS.
- RemCond removes and unregisters CASL conduit.
- CASLcopy copies program and any database files to user HotSync path.

Those are the basics of CASL.

PenRight! MobileBuilder

MobileBuilder from PenRight can be downloaded from www.penright.com/products-demo.htm, and a working sample in conjunction with Sybase SQL Anywhere Studio can be downloaded from www.sybase.com/testdrive. This product also supports multiple platforms and types of operating systems such as Palm, WinCE, Win16, Win32, and even DOS. Since the coverage of supported systems is large, you will need additional components to completely install all of the features.

For Windows CE-based devices:

- MobileBuilder relies on Microsoft eMbedded Visual Tools, which can be downloaded from www.microsoft.com/mobile/downloads/emvt30.asp, or Visual C++ 6.0 and the Windows CE Toolkit for Visual C++ 6.0.
- One or more supported Windows CE version 2.11 or greater SDKs supporting ARM, MIPS, SH3, SH4, or x86 processors.
- ActiveSync 3.1 or later.

For Palm-powered devices, the list of required products is a bit longer:

- PRC-Tools 2.0 or GCC 0.5.0 (both on the MobileBuilder CD).
- A PRC-Tools compatible Palm SDK.
- Palm Desktop (HotSync) 3.1 or later.
- ROM image (for emulator operation) that can be obtained by either downloading a ROM image from your own Palm-based device or by subscribing to *Palm Developer* at the Palm home site.

For DOS, DPMI, and Win16:

- Borland C++ 4.5, Borland C++ 5.0, or Visual C++ 1.5

For Win32:

- Borland C++ 4.5, Borland C++ 5.0, Visual C++ 5.0, Visual C++ 6.0

MobileBuilder documentation specifies that the minimum processor requirement for the system to run the previously listed components can be as low as an Intel 386. However, we would not attempt any productive development with MobileBuilder if the workstation does not satisfy the following requirements:

- Minimum Pentium II processor.
- Minimum of 64MB of RAM (128 is preferable).
- 300–400MB of disk space if all of the supported platforms will be used.
- Any of the following Windows operating systems: Windows 95, 98, NT, ME, 2000, and XP.

Now, for the last time in this chapter, we will build the "Hello World" application. As soon as we start the application, a screen similar to one shown in

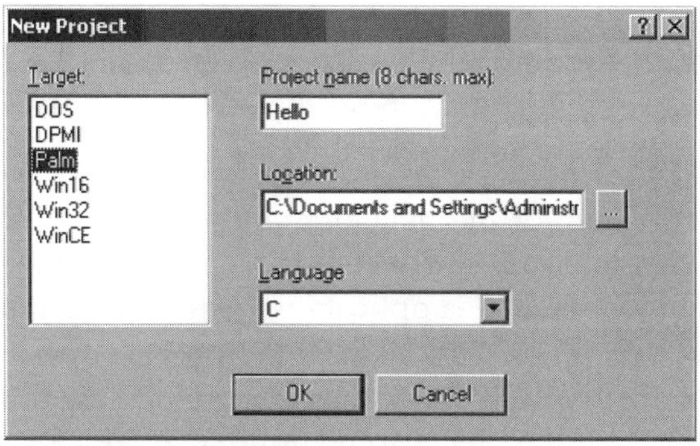

Figure 7.9 MobileBuilder project start.

Figure 7.9 will appear. You can see from this screen that coverage of the supported operating systems is by far the largest of all reviewed products, but the increased complexity affects the functionality, primarily in the field of database synchronization and data replication capabilities.

To begin, select the desired target operating system (in this case, that will be Palm), assign the project name and location for your project, and click OK. Your MobileBuilder will prepare an environment for the specific target platform of your choice, and the screen in Figure 7.10 will appear. This is the main part of MobileBuilder IDE and you can recognize more or less all of the components available across all of the tools.

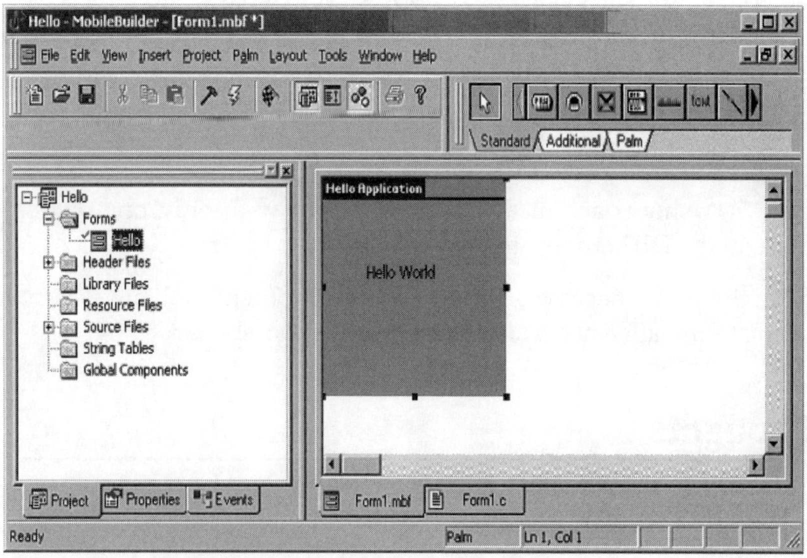

Figure 7.10 MobileBuilder "Hello World" application.

Where MobileBuilder is exceptional (apart from the operating system coverage) is the form presentation window where we are given the look and feel of the final product. You know immediately how the screen composition will look once deployed to a Palm-powered device.

The component that allows each of the development tools to communicate with different databases and determines what types of conduits are supported is what really defines usability of mobile applications. Databases populated or updated through MobileBuilder applications running on Palm OS and Windows CE devices can be transferred to the desktop using the tool's Palm conduit or Windows CE service provider. MobileBuilder currently supports file-level database transfer. This in effect means that only entire databases are transferred between the desktop and device as necessary. This is a large limitation since you, as developer, will often have to specify one-way or two-way synchronization, record filtering, and selective table-level synchronization. However, Sybase did an excellent job in integrating this product with their database engine. In fact, MobileBuilder is the recommended development environment for Sybase mobile applications.

DBArtisan

DBArtisan from Embarcadero is a great database administration tool and can be downloaded from the www.embarcadero.com site. While not a specific mobile application development tool, it does come in quite handy when administering one or a multitude of server databases. It focuses on server-based databases such as DB2, MS SQLServer, Sybase, and Oracle.

We've been fans of this tool for years now and find it extremely useful in managing and administering all the components of each of the databases. Since this project involved many types of database, this tool allowed us to quickly replicate database objects between distinct environments.

Figure 7.11 shows a typical screen shot, but with only one database called Testers in the DB2 environment.

Again, this is not necessarily a mobile development tool, but is very handy in administering all aspects of the server-based databases.

Final Thoughts

This chapter did not cover all of the available tools used for developing mobile applications. Some of the other tools available are Code Warrior (C++-based

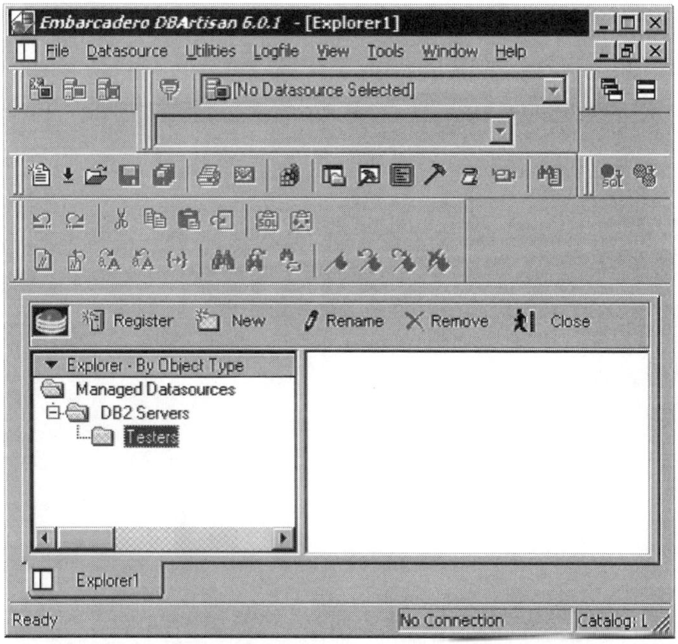

Figure 7.11 DBArtisan tool.

tool), NSBasic (Visual Basic-like syntax and IDE), and Visual Studio.NET with a fresh new approach to the development of the mobile applications.

It is important from the perspective of the developer to understand who your target audience is. Are they going to be Palm users, Windows CE users, or both? The answer to this question will help you select the tools you will use for development. An additional question you should ask yourself prior to selecting tools or starting development is, what type of database will you be talking to? For example, if your corporate database standard is MS SQL, you will not select IBM's DB2 Everyplace Mobile Application Builder.

The message is, choose your tools wisely. Happy programming.

CHAPTER 8

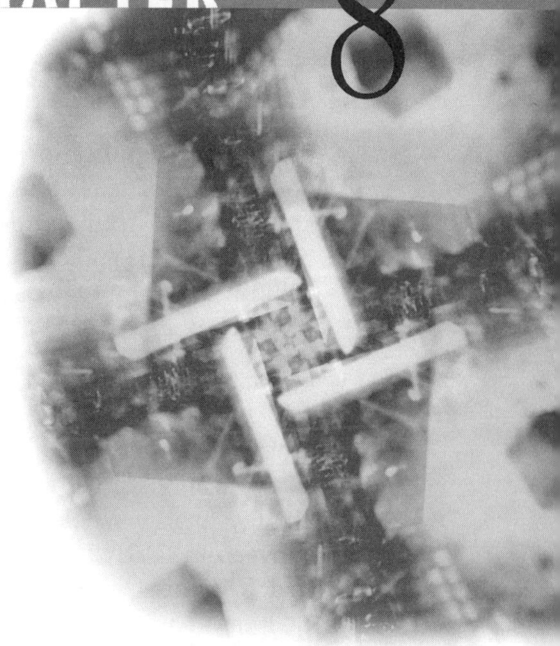

Palm's Database

This chapter is based on Palm OS powered devices, and hence the Palm PDB database structure. In order to test the applications provided in this chapter, you will need the following:

- A PC running the Windows operating system with Palm OS HotSync Manager installed.
- Visual Basic 6 with minimum SP 4.
- AppForge 2.1, (fully functional 30-day evaluation copy is available from) http://scripts.appforge.com/eval/afeval.asp).
- Palm OS-based device, or if you do not have one of the Palm or Visor devices, you can download the Palm Emulator from www.palmos.com/dev/tech/tools/emulator/.

NOTE

It is much easier to test and debug your application using the Emulator instead of using the device itself.

Once you are happy with the "look and feel" of your application, you can proceed with the deployment to the different types of Palm-powered devices. We'll talk more about this at the end of the chapter.

PDB Example: Fugitive Application

The sample application for this chapter is an imaginary fugitive-tracking application that could help police officers keep track of any wanted criminals still at large. We will guide you through the phases of building such an application using the standard Palm database (PDB), the native method for storing data on the Palm OS-based PDAs. There will be more about the structure of PDB in Appendix A, "Palm Conduits."

For data storage on the PC, we will use an Access database. As you will see in this chapter, any ODBC-compliant database can be used. We will build a fully functional application (you do not need to know C language in order to create an application for Palm devices) with a little help from our friends at AppForge. Finally, for the whole system to work there must be a conduit established between the PDA and your PC in order to exchange data and create database updates. This task is also a straightforward process because it is completely GUI-based and there is almost no need for coding whatsoever. The amount of coding is limited to the underlying Visual Basic controls, and the code itself supports almost all of the Visual Basic 6 syntax.

So, let's begin. The basic idea behind this application (as shown in Figure 8.1) is that each police officer will be equipped with a Palm-based PDA, which they will have to update daily with fresh data about dangerous criminals on the loose. In addition, PC databases will be updated from one central place daily, or we can configure our conduit in such a way as to talk to the central database directly.

To help you build this application, please take a moment to review application flow in Figure 8.8. You will notice immediately that our application will have a security login, allow browsing of records, search capability, a personal details screen, logoff capability, and an About info page. In addition, our application will have the capability to update records on Palm-based PDAs from a central database via the conduit.

NOTE

Before you even begin coding for Palm-based devices, you will need to apply with Palm for a Creator ID. This ID is unique for you and must be used, as you can see in Figure 8.6. The Creator ID must be four digits long, and it can be a combination of alpha and numeric characters. You can apply online for a Creator ID at http://dev.palmos.com/creatorid/reg.cfm.

To start, we have to first create an Access database. As you can see, we will have only two tables created for the Mobile Officer database. The Security table will be used to validate login information and hold users' personal data.

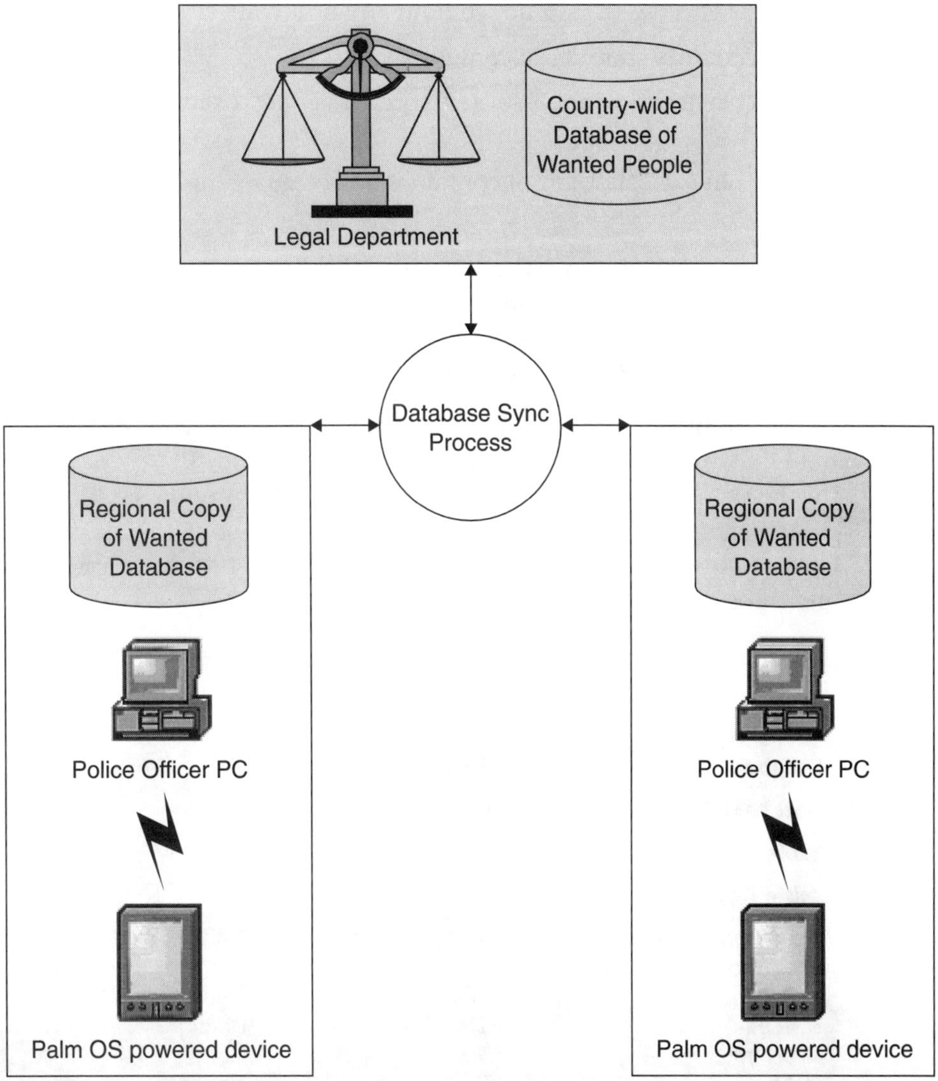

Figure 8.1 Overall sample application idea.

The Wanted table will be used to keep a list of all the records that are downloaded from your PC. Table layouts are shown in Figure 8.2. Some rules will help you successfully design, deploy, and sync data:

- It is often the case that your tables have some type of primary key, usually to ensure data integrity and uniqueness of the records. If this is the same for your tables, then use the same key as your sync key.

- Your sync key should not contain any null values.
- Field names should not contain spaces.
- Field names cannot use keywords that are used by the host database engine.
- The sync key should not contain fields of type Single or Boolean.

If you are planning to use an enterprise class database (MSSQL, Oracle, or something similar), you can still use MS Access for your development. When you are ready for production testing, you can simply upsize your database to the more serious database engine.

At this point, you should have created a database on your desktop computer, but what about the selection of the database on your PDA? As you will see in this book, all of the major database vendors do provide small memory footprint versions of their "big siblings" databases, but that is not your only choice. Our choice for this chapter is PDB format, the native format of storing and transferring data to your PDA.

Database Title: **Mobile Officer.**

Database Path: **C:\PolAlert.mdb**

Table Security **Created** 12/30/2001 **Modified** 1/5/2002 **Fields** 5

Table Description

Field Name	Indexed	Field Type	Size	Default Value	Reqd
BadgeNo	Primary Key	Long	4	0	No
FirstName		Text	50	None	No
LastLogin		Date/Time	8	Date()	No
LastName		Text	50	None	No
PWD	Duplicates OK	Text	8	None	No

Table Wanted **Created** 12/30/2001 **Modified** 1/6/2002 **Fields** 6

Table Description

Field Name	Indexed	Field Type	Size	Default Value	Reqd
Chars		Text	150	None	No
Dangerous		Yes/No	1	None	No
FirstName		Text	50	None	No
LastName		Text	50	None	No
OnTheRun		Yes/No	1	None	No
RecordNo	Primary Key	Long	4	None	No

Figure 8.2 Access database structure.

PDB Database Components

We know that on the desktop PC, databases are stored (at least initially) on the file system, but what about Palm-powered devices? Databases and programs are held in memory, which is one of the major reasons why the amount of memory on the PDA is so important. PDB file format is sequential and has the following major components (sections): database header, record list, application information block (optional), sort information block (optional), and raw record data. These sections can be placed into three groups as you can see in Table 8.1.

The Palm database header has the structure (in fact, "C" Language structure) shown in Listing 8.1.

```
#define dmDBNameLength    32
typedef struct {
UInt8 name[dmDBNameLength];
UInt16 attributes;
UInt32 creationDate;
UInt32 modificationDate;
UInt32 lastBackupDate;
UInt32 modificationNumber;
LocalID appInfoID;
LocalID sortInfoID;
UInt32 type;
UInt32 creator;
UInt32 uniqueIDSeed;
RecordListType recordList;
} DatabaseHdrType;
```

Listing 8.1 PDB header structure.

The explanation for each of these fields can be found in Table 8.2.

Table 8.1 PDB Database Structure

STANDARD HEADER INFORMATION	DATABASE HEADER LIST OF RECORD ENTRIES
Application-specific information (this is optional)	Application information block Sort information block
Application records	Sequence of record data

Table 8.2 PDB Header Details

name	Field that contains the name of the Palm database and is a 32-byte long, null-terminated string containing the name of the database on the Palm-powered handheld.
attributes	The attribute flags for the database.
version	Version of the database layout. This information is application specific.
creationDate	The creation date of the database. This field is a calculated field and represents the number of seconds since January 1, 1904 at 12:00 A.M.
modificationDate	The most recent modification of the database represented (like previous field) in the number of seconds since January 1, 1904 at 12:00 A.M.
lastbackupDate	The date and time of the most recent backup since January 1, 1904 at 12:00 A.M.
modificationNumber	This field stores the modification number of the database.
appInfoID	This field represents the offset from the beginning of the database header to the start of the appInfo segment.
sortInfoID	This field represents the offset from the beginning of the database header to the start of the sortInfo segment.
type	The database type identifier.
creator	The database creator identifier. This is one of the reasons why you must apply for a developer ID with 3Com.
uniqueIDSeed	As its name indicates, this field is used to generate unique records on the Palm database while the database is loaded into the PDA.
recordList	A list of the records in the database.

The Palm database header ends with a record list. Each of the entries in the record list defines a single record. The "C" structure in Listing 8.2 represents a Palm database record list.

Each of the fields has specific meaning as you can see in Table 8.3.

```
typedef struct {
LocalID nextRecordListID;
UInt16 numRecords;
UInt16 firstEntry;
} RecordListType;
```

Listing 8.2 PDB record list.

Table 8.3 PDB Record List

nextRecordList	Used to point to the next record list. This field is almost always 0, since there is almost never more than one record list.
numRecords	Number of entries in the record list.
firstEntry	This field represents the start of an array of record entry structures.

Records (or record data) are stored in a PDB as a block of adjoining records. The beginning of each record is stored as a local offset value in the database header (remember the field recordList?). If you would like to learn more about PDB format (as well as other native Palm formats such as PRC and PQA), please visit www.palmos.com/dev/tech/docs/ where you can find the latest Palm documentation. One of the advantages of the selected toolkit for this chapter is that it supports PDB files and allows you to open them on the PC. We will talk about data synchronization in more detail later in this chapter.

AppForge 2.1 Setup and Installation

After deciding the type of the database to use on the desktop and Palm, your next step would be to download and install AppForge on your PC, and the Booster component on your Palm OS-powered device (you can find the download site at the beginning of this chapter). Installation is straightforward and can be performed very quickly, as long as you have VB 6 already installed on your system. The entire installation process does not take more than five minutes. The first part of the installation involves AppForge, which a is cross-platform development tool that enables you to create one application that can be deployed to two major PDA operating systems on the market today. The second part of the installation involves Booster. What is Booster? In simple terms, it is set of three components (AFCore, pCom, and ByteStreamVM) that are in charge of core AppForge services, component management, and interpretive execution of the code. Because AppForge is a cross-platform tool, there is version of the Booster component for Palm- and CE-powered devices. For the Palm-based devices, Booster (in the form of BOOSTER.PRC) is a free component; however, if you need Booster for CE, you will have to pay for the license.

In order to upload (install) your application to the PDA, AppForge will create a PRC file, which is nothing more than the compiled Palm application from the code that you have created on the VB side. This file will be copied into the Palm Installation directory. If you have selected any dependent PDB files, they will be also copied in the Installation Palm installation directory. The entire project is uploaded to the device during the next HotSync.

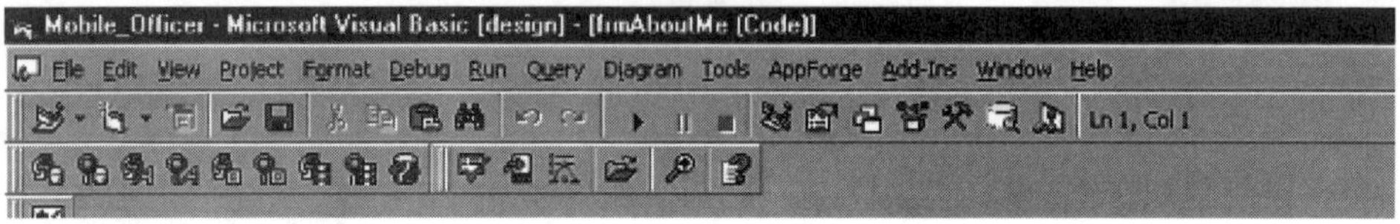

Figure 8.3 VB toolbar.

Now, let's go back to the development of the application. Once you have downloaded, installed, and registered your evaluation version of AppForge, you will be able to see an additional toolbar available to you on the VB IDE (see Figure 8.3).

Each of these additional toolbar icons shown in Figure 8.4 represents tasks that are almost separate applications.

As you can see in Figure 8.5, the same functionality can be accessed from the AppForge drop-down menu in your VB user IDE. It is really a question of personal preference as to which method you will use the most. We personally find toolbar icons to be a much more efficient way to "move around" AppForge. Intuitiveness is one of the major features, and it enables even the novice developer to quickly (and successfully) build and deploy portable applications.

Once you have your source database created, you will have to use the Database converter tool to create a PDB format database. Because PDB files are not relational and they have separate files for each table structure, conversion will have to take place one table at a time. In our example, this will require you to use the Database Converter tool twice consecutively—once for the Security table and the second time for the Wanted table. In addition, this is one of the places where you will have to provide your unique "Creator ID." Please do not use arbitrary

Database Converter Tool	PDB Database Viewer
Font Converter Tool	Font Viewer Tool
Graphic Converter Tool	Graphic Viewer Tool
Movie Converter Tool	Movie Viewer Tool
Universal Conduit Configuration Tool	Compile and Validate Tool
Deploy to Device Tool	AppForge Settings
Open New Project	Zoom Level Tool
AppForge Users Guide	

Figure 8.4 AppForge toolbar.

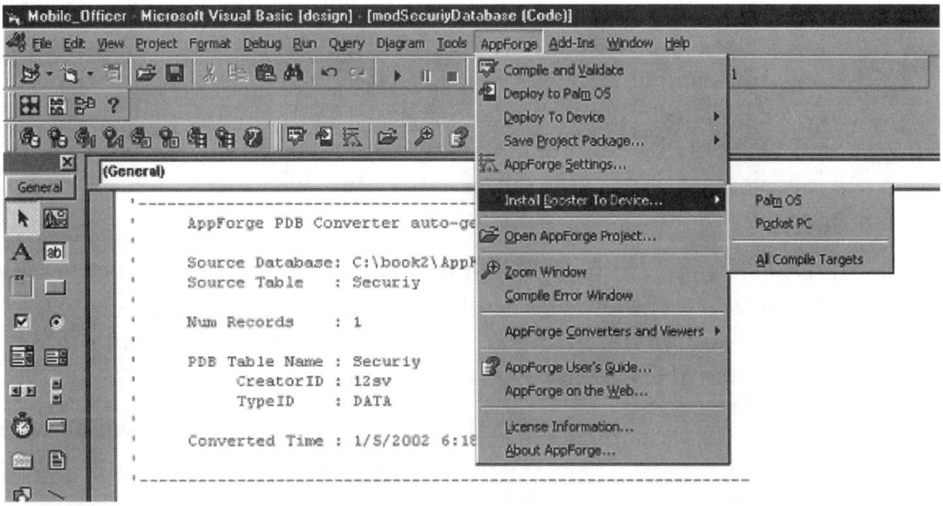

Figure 8.5 AppForge drop-down menu.

strings here, because that kind of action could have unwanted effects. The screenshot for this process is shown in Figure 8.6. Now, remember for a second what the structure of PDB files is. During the conversion process, the major parts of the PDB file will be created (headers and records). Depending on the number of records in your tables, this process can take some time to complete.

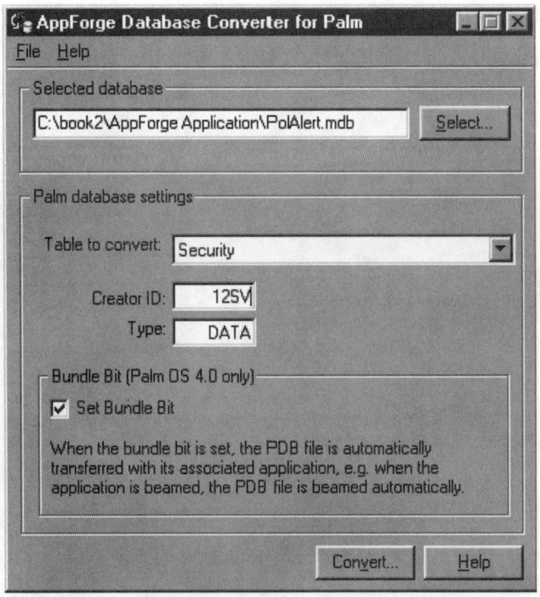

Figure 8.6 AppForge Database Converter.

One of the differences between versions 2.0 and 2.1 of AppForge is that in version 2.1 (see Figure 8.6), there is a check box called "Set Bundle Bit." This bit is available on Palm-powered devices that have version 4.0 of Palm OS or higher. By checking this box you ensure that the database will be delivered to the Palm-powered device at the same time the PRC file is delivered. In the previous version, you had to manually define files that would be transferred to the Palm device.

It is also important to note that not all field types can be used (on the Access side) if you want to successfully use the data conversion tool. Table 8.4 lists field types that are used by Access tables and corresponding field types in PDB format, if they exist.

Figure 8.7 shows the confirmation process that will occur while the database or databases have been converted and prepared for transfer to your portable device during the next Hot-Sync process.

An important feature of AppForge is that it will help you create a conduit to your desktop (Access) or Enterprise-wide database (MSSQL, Oracle or DB2) via a set of very intuitive GUI interfaces. As long as you have an appropriate ODBC DSN source, this should not be hard to configure. First, you have to create a new Universal Conduit Descriptor (UCD).

Table 8.4 Access and PDB Field Type Comparison

ACCESS DATA TYPES	SUPPORTED BY APPFORGE 2.1	PDB DATA TYPES
Text	✔	String
Memo	✔	String
Number, Byte	✔	Byte
Number, Integer	✔	Integer
Number, Long Integer	✔	Long
Number, Single	✔	Single
Number, Double	✔	Double
Number, ReplicationID	✘	No equivalent
Number, Decimal	✘	No equivalent
Date/Time	✔	Date
Currency	✔	Currency
AutoNumber	✔	Long
Yes/No	✔	Boolean

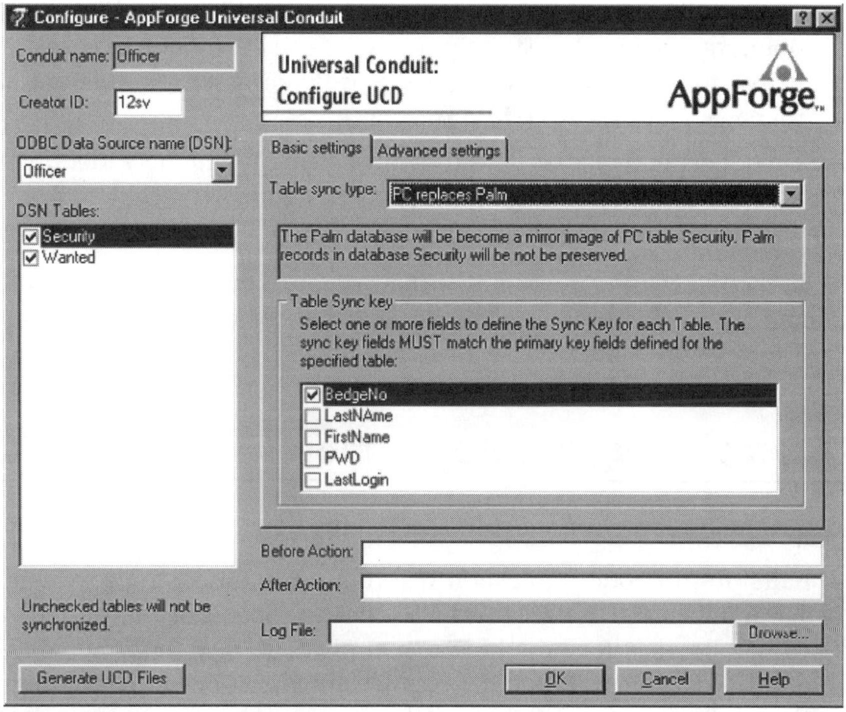

Figure 8.7 UCD configuration.

NOTE

The Conduit Creation tool will not be available unless you purchase the AppForge license.

Please note the significance of defining the Sync Key field in AppForge (see Figure 8.7). A sync key is associated with a table, and defines the set of fields that determine the identity of a record. The sync key is used to compare records between a Palm database and a corresponding host table.

Some suggestions for selection of the key fields include:

- If your database contains a primary key, you must select it as your sync key.
- You should not use keywords such as "String" or "Integer" as field names.
- Fields that could be set to a null value should not be selected for key fields.
- There must be no spaces in any of the fields within a sync key. For example, "Last Name" would be illegal, but "LastName" would be the correct selection for the field name.

- The sync key should not contain Single or Boolean field types. Although doing so is allowed, floating-point types can be difficult to check for equality, while Boolean types might sort differently across different Database Management Systems (DBMS).
- Do not depend on case sensitivity in a string to uniquely identify a record, as case is ignored during sync key comparison.

If defined, this field must match the primary key (or keys) defined for the source table. This process must be repeated for each database that will be transferred to the PDA.

PDB Example: Fugitive Application

Our sample application will be based on the flow chart show in Figure 8.8.

The first form in our application will require an officer to provide his or her badge number and assigned user ID. In the context of this application, this form will be used to show how one application or data presented with the application can be protected. Listing 8.3 and Figure 8.9 represent Login forms as they will be seen in VB and a Palm-powered device, respectively. You can almost say that once you start an AppForge project within your VB session, you have a What You See Is What You Get (WYSIWYG) interface. The size of the form created in VB is proportional to the real-life look and feel of the form, once it is deployed to the device.

Now, let's "walk" through the code of this form. During the Form_Load event we will open the connection, or to be more precise in this case, open the database by invoking the OpenSecurityDatabase procedure. This procedure is part of the VB module (you can see details of this module in Listing 8.10) that is created (generated) during the conversion to PDB format by AppForge. By doing this, AppForge is ensuring consistency and helping developers. You can create database open statements manually if you prefer, but it is much easier to leave this task to AppForge.

One great thing about AppForge as an IDE for PDA-based development is that it offers multiple layers of emulation and testing:

- Almost WYSIWYG GUI interface that (apart from the improper background color) shows the PDA form in the size that is very close to the original.
- Palm Emulator where we can load an application and get results in almost exactly the same look and feel of the application that is delivered to the real Palm-based device.

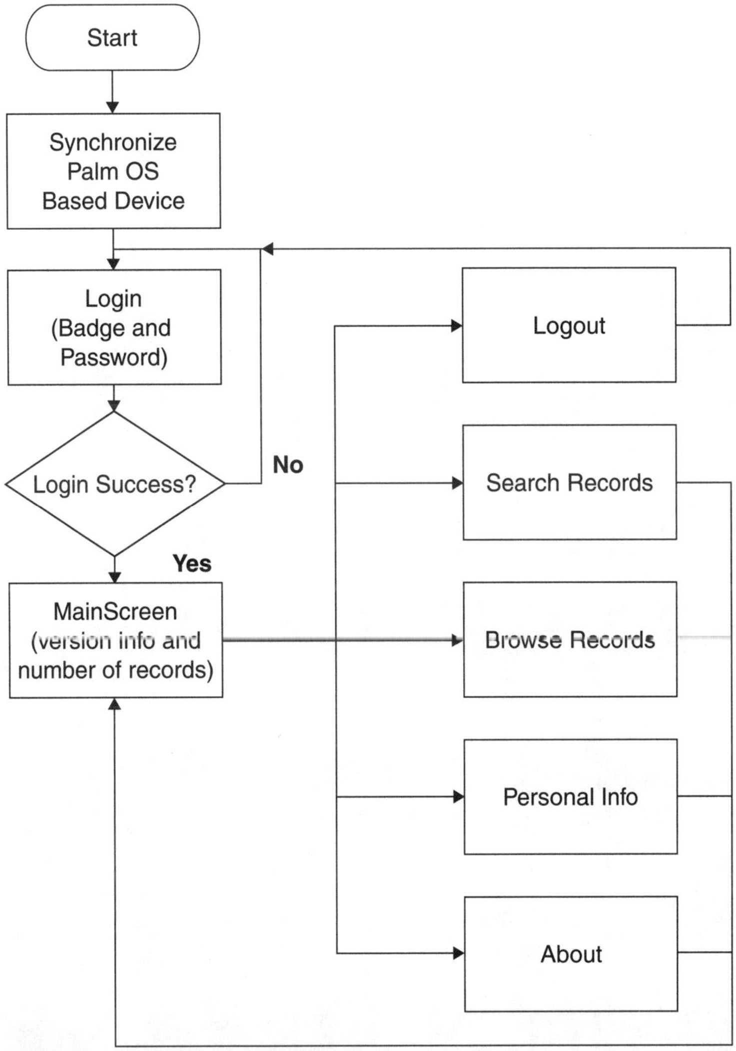

Figure 8.8 Fugitive application flow.

You can see here how you can build and test your application. Only when you are completely convinced that all the components are ready to be propagated to the devices should you proceed with deployment to the physical device (see Figure 8.9).

The code in Listing 8.3 has a very simple task: Check data entered in two form controls (AFTextBox1 and AFTextBox2). Every time the form is loaded, two

```
1. '-------------------------------------------------------------
2. '     Login Form for Mobile Officer demo application
3. '
4. '          Form Name : frmLogin
5. '          CreatorID : 12sv
6. '          TypeID    : PRG
7. '
8. '       Created Time : 1/5/2002 6:19:42 AM
9. '-------------------------------------------------------------

10. Option Explicit
11. Dim strBadge As String
12. Dim strpass As String

13. Private Sub AFButton1_Click()
14. Dim MyRecord As tSecurityRecord
15. PDBMoveFirst (dbSecurity)
16. strBadge = Trim(AFTextBox1.Text)
17. strpass = Trim(AFTextBox2.Text)
18. If PDBNumRecords(dbSecurity) > 0 Then
19. ReadSecurityRecord MyRecord
20. If strBadge = Trim(MyRecord.BadgeNo) And strpass =
       Trim(MyRecord.PWD) Then
21. AFLabel3.Caption = ""
22. AFTextBox1.Text = ""
23. AFTextBox2.Text = ""
24. Me.Hide
25. frmMain.Show
26. Else
27. AFTextBox1.Text = ""
28. AFTextBox2.Text = ""
29. AFLabel3.Caption = "Invalid Login"
30. AFTextBox1.SetFocus
31. End If
32. End If

33. End Sub

34. Private Sub Form_Activate()
35. AFTextBox1.SetFocus
36. End Sub

37. Private Sub Form_Load()
38. AFTextBox1.Text = ""
39. AFTextBox2.Text = ""
40. AFLabel3.Caption = ""
41. OpenSecurityDatabase
42. End Sub
```

Listing 8.3 frmLogin code.

previously mentioned fields are “cleaned up” (lines 37 through 42 in Listing 8.3). Once data has been entered, execute function AFButton1_Click (lines 13 through 33 in our code listing) by clicking on the button AFButton1. Applica-

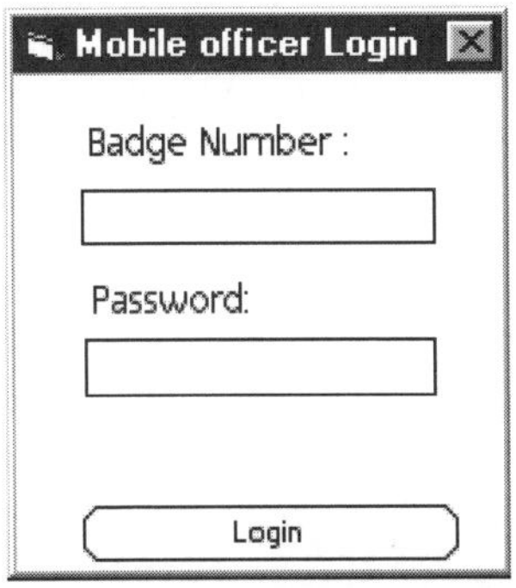

Figure 8.9 Form created by executing VB code.

tion logic will allow for only one user to be defined in the database, and for that reason it will simply position the record pointer to the first field in the database by executing code on line 15 (PDBMoveFirst [dbSecurity]). It is easy to imagine that adding complete search capability during the login process, thus allowing multiple users to be defined in the database, can enhance this part of the form (see Figure 8.10).

In case a user enters an incorrect user ID or password, form fields will be cleared from any text and the user will be forced to enter information again. Our user is in once the correct information has been entered. Note that the database is opened during the Form_Load event (line 41 in Listing 8.3), and that each database has to be opened separately in order to access stored records.

Based on our application flow chart (Figure 8.8), the next step in the demo application for this chapter would be to create a form that will serve as a focal point for other options to be selected and to show how many fugitives we have in our database. No wonder the name given for the form in this case would be "frmMain."

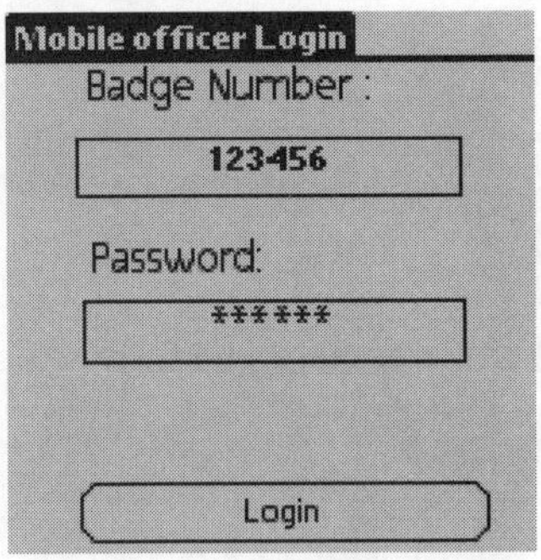

Figure 8.10 Login form—Palm screenshot.

The first thing we would like our user to see is the menu bar with a drop-down box that will present the user with available options and number of records. You can see the screenshot for this form (taken from the Palm Emulator) in Figure 8.11. If you look at the code in Listing 8.4, you will notice that there is no code for the menu itself. This is because we have used VB's built-in "Menu Editor." Since this chapter is not about Visual Basic, we will not go into too many details about the VB itself.

```
'-----------------------------------------------------------
'      Main Form for Mobile Officer demo application
'
'           Form Name : frmMain
'           CreatorID : 12sv
'           TypeID    : PRG
'
'        Created Time : 1/5/2002 6:19:42 AM
'-----------------------------------------------------------

Option Explicit

Private Sub About_Click()
Me.Hide
frmAboutMe.Show
End Sub

Private Sub Form_Load()
OpenWantedDatabase
AFLabel1.Caption = PDBNumRecords(dbWanted)
End Sub

Private Sub Inet_Click()
Me.Hide
frmBrowRec.Show
End Sub

Private Sub Logout_Click()
Me.Hide
frmLogin.Show
End Sub

Private Sub PersSet_Click()
Me.Hide
frmPersOpt.Show
End Sub

Private Sub Search_Click()
Me.Hide
frmSearch.Show
End Sub
```

Listing 8.4 frmMain code.

As you can see in Listing 8.4, during the Form_Load event we will open the Wanted database and execute AppForge PDBNumRecords([database name]), which will give us the total number of records in the database and present that information on the display of the device (see Figure 8.11). This is achieved with the simple code on lines 16 and 17.

All remaining code in this form is used to "connect" to appropriate forms as per our application flow definition shown in Figure 8.8.

At this point, our user will be able to take multiple actions. As with any other database application, end users will be able to browse, search, and append records. The only thing that we will leave to AppForge Universal Conduit is synchronization of the data, which is, in our case, one way for the Wanted database and two ways for the Security database.

So, let's start with the data browsing. As always, the first event executed in this form is Form_Load. We can see on lines 28 through 32 in Listing 8.5 that by executing AppForge function PDBSetSortFields, we will set sort order to the LastName field in the Wanted database. Following the line of code will position the record pointer to the first field in the database. Line 31 will then invoke the procedure that will actually retrieve the data from the database and assign each of the form objects (text boxes and check boxes) appropriate values from it (lines 33 through 50).

Once we have data on the form, the only parts of the browsing the end user would execute would be to move the record pointer up and down in the database and to exit the form. Lines 16 through 21 in Listing 8.5 will move the record pointer to the next field in the database and refresh the form controls with the new data. A similar action (but only for the movement in the direction of the previous record) will be achieved with lines 22 through 27. The only detail left is to provide the user with a method to close this form and return to our main form,

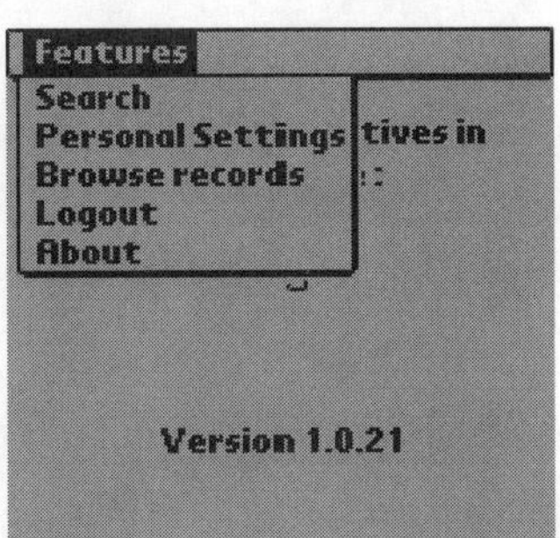

Figure 8.11 Application main menu.

```
'---------------------------------------------------------------
'      Record Browser Form for Mobile Officer demo application
'
'           Form Name : frmBrowRec
'           CreatorID : 12sv
'           TypeID    : PRG
'
'         Created Time : 1/5/2002 6:19:42 AM
'---------------------------------------------------------------

Option Explicit
Dim MyRecord As tWantedRecord

Private Sub AFButton1_Click()

Me.Hide
frmMain.Show
End Sub

Private Sub AFButton2_Click()

PDBMoveNext (dbWanted)
If Not PDBEOF(dbWanted) Then
DisplayData
End If
End Sub

Private Sub AFButton3_Click()

PDBMovePrev (dbWanted)
If Not PDBBOF(dbWanted) Then
DisplayData
End If
End Sub


Private Sub Form_Load()
PDBSetSortFields dbWanted, LastNAme1_Field
PDBMoveFirst (dbWanted)
DisplayData
End Sub

Public Sub DisplayData()

If PDBNumRecords(dbWanted) > 0 Then
ReadWantedRecord MyRecord
AFTextBox1.Text = MyRecord.LastName
AFTextBox2.Text = MyRecord.FirstName
If MyRecord.OnTheRun Then
AFCheckBox2.Value = afCheckBoxValueChecked
Else
AFCheckBox2.Value = afCheckBoxValueUnchecked
End If
```

Listing 8.5 frmBrowRec code.

```
43. If MyRecord.Dengerous Then
44. AFCheckBox1.Value = afCheckBoxValueChecked
45. Else
46. AFCheckBox1.Value = afCheckBoxValueUnchecked
47. End If
48. AFTextBox5.Text = MyRecord.Chars
49. End If
50. End Sub

51. Public Sub initialize()

52. loadRec = PDBReadRecord(dbWanted, VarPtr(RecToShow))
53. End Sub
```

Listing 8.5 *(Continued)*

which will allow the user to perform a new action. This is done on lines 12 through 15. Figure 8.12 shows how this form would look in real life.

Now, what kind of database application would this be if we did not show you how to update the records on the Palm-powered device? Let's go to the personal options part of the application where we will be able to update the security password. This form will enable our user to view personal details (last name, first name, badge number, and creation date), and also allow modification of the password field. Again, the first event that is executed upon the start of the application is Form_Load (lines 24 through 35 in Listing 8.6). Once all the screen fields have been populated from the Security table, the only field that users will be able to change the values of is Password.

Figure 8.12 Browse fugitive records.

```
1. '---------------------------------------------------------------
2. '   Personal Options Form for Mobile Officer demo application
3. '
4. '          Form Name : frmPersOpt
5. '          CreatorID : 12sv
6. '          TypeID    : PRG
7. '
8. '       Created Time : 1/5/2002 6:19:42 AM
9. '---------------------------------------------------------------

10. Option Explicit
11. Dim MyRecord As tSecurityRecord
12. Dim WriteToDB As Long

13. Private Sub About_Click()
14. frmAboutMe.Show
15. End Sub

16. Private Sub AFButton1_Click()
17. MyRecord.PWD = AFTextBox4.Text
18. PDBEditRecord dbSecurity
19. WriteToDB = PDBWriteRecord(dbSecurity, VarPtr(MyRecord))
20. PDBUpdateRecord dbSecurity
21. Me.Hide
22. frmMain.Show
23. End Sub

24. Private Sub Form_Load()
25. PDBMoveFirst (dbSecurity)
26. If PDBNumRecords(dbSecurity) > 0 Then
27. ReadSecurityRecord MyRecord
28. AFLabel6.Caption = MyRecord.BadgeNo
29. AFLabel6.Refresh
30. AFTextBox2.Text = MyRecord.LastName
31. AFTextBox3.Text = MyRecord.FirstName
32. AFTextBox4.Text = MyRecord.PWD
33. AFLabel5.Caption = MyRecord.LastLogin
34. End If
35. End Sub
```

Listing 8.6 frmPersOpt code.

Thanks to functions delivered by AppForge, this task can be accomplished in a few lines of code. Once the user presses the Confirm Changes button, code on lines 16 through 23 will take care of the edit-assign value-update sequence for the password field (if the user has made any changes to this field). The look of the screen that is produced (with execution of the code in Listing 8.6) on the screen of a Palm-based device is shown in Figure 8.13.

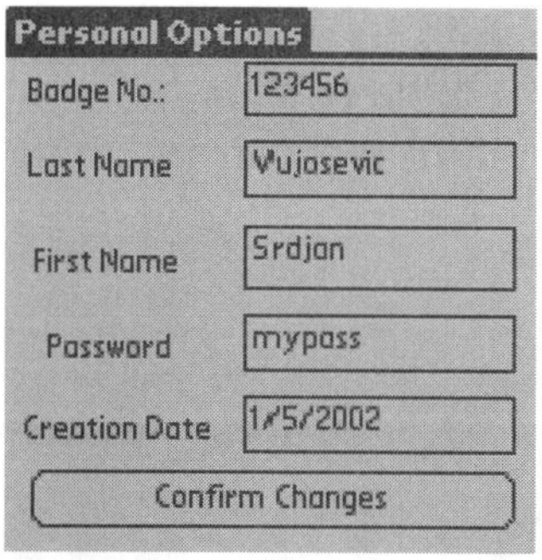

Figure 8.13 Personal options screen.

For the purposes of this chapter, we have created a database that has only three records. In reality, you will have more than that, and one of the features you will have to provide for your customers is search. In this example, we will provide this feature via two forms, one that will allow your user to select search criteria and enter search keywords, and one that will present the user with the search details. As you can see in Figure 8.14, our user will have to select search field type in the keyword field and click the Search button.

If you examine the program (represented by Listing 8.7), you will notice that it is built from four major parts:

- Form_Load subroutine (lines 54 through 66) that populates the search field combobox.
- A subroutine that will perform search, after the user presses the Search button (lines 18 through 45). This subroutine will also populate the Results list box with all of the search results.

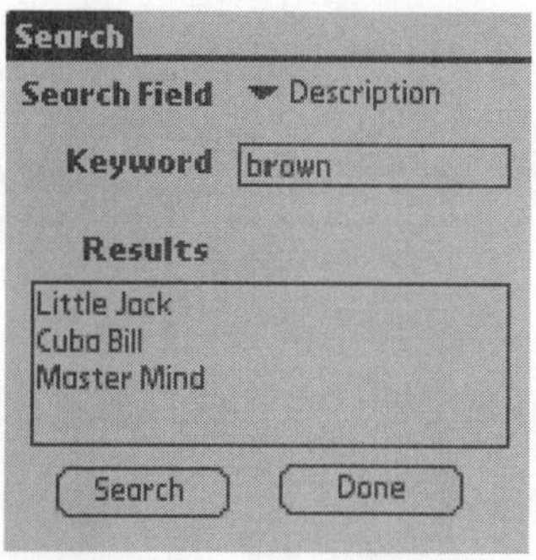

Figure 8.14 Search screen.

- A subroutine that will load data for the record that the user has selected from the list of possible matches (lines 67 through 76).
- Subroutines that will close the form and return to the main form.

```
1. '-------------------------------------------------------------
2. '     Search Form for Mobile Officer demo application
3. '
4. '          Form Name : frmSearch
5. '          CreatorID : 12sv
6. '          TypeID    : PRG
7. '
8. '       Created Time : 1/5/2002 6:19:42 AM
9. '-------------------------------------------------------------

10. Option Explicit

11. ' Define variables that will be used in the form
12. Private arr_lSearchFldIdx(3) As Long
13. Private lSelectedSearchSectionIdx As Long
14. Private SearchFor As String
15. Private bFindSuccessful As Boolean
16. Private bDoneFinding As Boolean
17. Private sThisFieldValue As String


18. Private Sub AFButton1_Click()
19. If txtKeyword.Text <> "" Then
20. Dim FieldIdx As Long
21. SearchFor = txtKeyword.Text
22. lstResults.Clear
23. FieldIdx = arr_lSearchFldIdx(lSelectedSearchSectionIdx)

24. bFindSuccessful = False
25. bDoneFinding = False

26. PDBMoveFirst dbWanted

27. While bDoneFinding = False
28. sThisFieldValue = modWantedDatabase.fnGetFieldByIdx(FieldIdx)

29. PDBGetField dbWanted, FieldIdx, sThisFieldValue

30. If InStr(1, LCase(sThisFieldValue), LCase(SearchFor)) Then
31. lstResults.AddItem modWantedDatabase.fnGetFieldByIdx(tWantedData
       baseFields.LastNAme1_Field) & " " & modWantedDatabase
       .fnGetFieldByIdx(tWantedDatabaseFields.FirstName_Field), -1
32. lstResults.ItemData(lstResults.ListCount - 1) =
       PDBRecordUniqueID(dbWanted)
33. End If
34. PDBMoveNext dbWanted
```

Listing 8.7 frmSearch code.

```
If PDBEOF(dbWanted) Then
bDoneFinding = True
End If
Wend

If lstResults.ListCount > 0 Then
bFindSuccessful = True
Else
lstResults.AddItem "(No matches found.)", -1
End If
End If
End Sub

Private Sub AFButton2_Click()
Me.Hide
frmMain.Show
End Sub

Private Sub cmbSelection_Click()
lSelectedSearchSectionIdx = cmbSelection.ListIndex 'Set search index to dropdown index
txtKeyword.SetFocus    'Set focus back to search word field
End Sub

Private Sub Form_Load()
cmbSelection.Clear
cmbSelection.AddItem "Last Name"
cmbSelection.AddItem "First Name"
cmbSelection.AddItem "Description"
cmbSelection.AddItem "Fugitive"
cmbSelection.ListIndex = 0

arr_lSearchFldIdx(0) = tWantedDatabaseFields.LastNAme1_Field
arr_lSearchFldIdx(1) = tWantedDatabaseFields.FirstName_Field
arr_lSearchFldIdx(2) = tWantedDatabaseFields.Chars_Field
arr_lSearchFldIdx(3) = tWantedDatabaseFields.OnTheRun_Field

lSelectedSearchSectionIdx = 0
End Sub

Private Sub lstResults_Click()
If bFindSuccessful = True Then
PDBFindRecordbyID dbWanted, (lstResults.ItemData(lstResults.ListIndex))

loadRec = PDBReadRecord(dbWanted, VarPtr(RecToShow))
frmView.initialize
frmView.Show
frmView.Refresh
Me.Hide
End If
End Sub
```

Listing 8.7 *(Continued)*

Once the user selects one of the records from the list of available (possible) matches (in this case, that will be Little, Jack), a form with the code in Listing 8.8 will be executed. In this case, nothing fancy will be in the code. We just want to present the user with a record that he or she selected.

This form (Figure 8.15) is view only, and the user cannot change any data.

If you want to show static information to your user, the process is straightforward; one label and one command button will do the job. This is presented by the About form in our example. (Listing 8.9 shows the code, and Figure 8.16 shows the resulting screen.)

One of the features that you will have to offer (especially since we have login functionality) is logout. In our case, that would be simply the option from the

```
'-------------------------------------------------------------
'      Search View Form for Mobile Officer demo application
'
'            Form Name : frmView
'            CreatorID : 12sv
'            TypeID    : PRG
'
'         Created Time : 1/5/2002 6:19:42 AMa
'-------------------------------------------------------------

Option Explicit
Dim MyRecord As tWantedRecord

Private Sub AFButton1_Click()
Me.Hide
frmMain.Show
End Sub

Public Sub initialize()
AFTextBox1.Text = RecToShow.LastName
AFTextBox2.Text = RecToShow.FirstName
If RecToShow.OnTheRun Then
AFCheckBox2.Value = afCheckBoxValueChecked
Else
AFCheckBox2.Value = afCheckBoxValueUnchecked
End If
If RecToShow.Dengerous Then
AFCheckBox1.Value = afCheckBoxValueChecked
Else
AFCheckBox1.Value = afCheckBoxValueUnchecked
End If
AFTextBox5.Text = RecToShow.Chars
End Sub
```

Listing 8.8 frmView code.

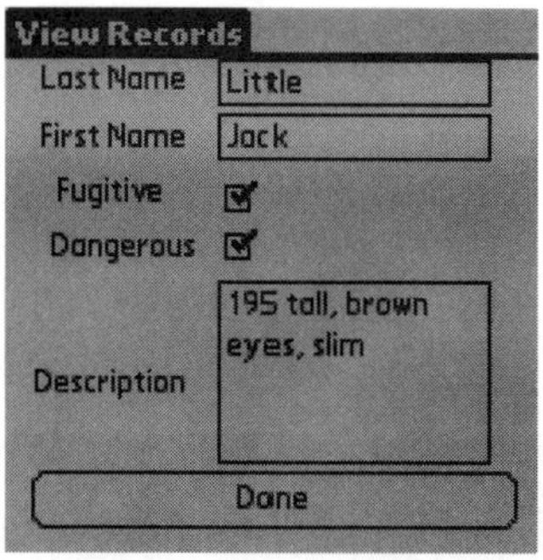

Figure 8.15 View search results screen.

```
'---------------------------------------------------------------
'      About Form for Mobile Officer demo application
'
'          Form Name : frmAboutMe
'          CreatorID : 12sv
'          TypeID    : PRG
'
'       Created Time : 1/5/2002 6:19:42 AM
'---------------------------------------------------------------

Option Explicit

Private Sub AFButton1_Click()
Me.Hide
frmMain.Show
End Sub
```

Listing 8.9 frmAboutMe code.

Figure 8.16 About form screen.

menu on the main form that will point the user back to the Login form (Figure 8.10), which would complete the circle.

Listings 8.10 and 8.11 are generated (except small details that we will point to you) by AppForge during the database conversion phase of the project.

```
'---------------------------------------------------------------
'      AppForge PDB Converter auto-generated code module
'
' Source Database: C:\book2\AppForge Application\PolAlert.mdb
'      Source Table    : Security
'
'      Num Records     : 1
'
'      PDB Table Name : Security
'            CreatorID : 12sv
'            TypeID    : DATA
'
'      Converted Time : 1/5/2002 6:18:20 AM
'
'---------------------------------------------------------------

Option Explicit

' Use these constants for the CreatorID and TypeID
Public Const Security_CreatorID As Long = &H31327376
Public Const Security_TypeID As Long = &H44415441

' Use this global to store the database handle
Public dbSecurity As Long

' Use this enumeration to get access to the converted database Fields
Public Enum tSecurityDatabaseFields
BedgeNo_Field = 0
LastNAme_Field = 1
FirstName_Field = 2
PWD_Field = 3
LastLogin_Field = 4
End Enum

Public Type tSecurityRecord
BedgeNo As Long
LastName As String
FirstName As String
PWD As String
LastLogin As Date
End Type
Public Function OpenSecurityDatabase() As Boolean
' Open the database
```

Listing 8.10 AppForge-generated code for the Security table.

```
#If APPFORGE Then
dbSecurity = PDBOpen(Byfilename, "Security", 0, 0, 0, 0, afModeReadWrite)
#Else
dbSecurity = PDBOpen(Byfilename, App.Path & "\Security", 0, 0, 0, 0, afModeReadWrite)
#End If

If dbSecurity <> 0 Then
'We successfully opened the database
OpenSecurityDatabase = True

Else
'We failed to open the database
OpenSecurityDatabase = False
#If APPFORGE Then
MsgBox "Could not open database - Security", vbExclamation
#Else
MsgBox "Could not open database - " + App.Path + "\Security.pdb" +
    vbCrLf + vbCrLf + "Potential causes are:" + vbCrLf + "1. Database
    file does not exist" + vbCrLf + "2. The database path in the
    PDBOpen call is incorrect", vbExclamation
#End If
End If

End Function

Public Sub CloseSecurityDatabase()
' Close the database
PDBClose dbSecurity
dbSecurity = 0
End Sub

Public Function ReadSecurityRecord(MyRecord As tSecurityRecord) As
    Boolean
ReadSecurityRecord = PDBReadRecord(dbSecurity, VarPtr(MyRecord))
End Function

Public Function WriteSecurityRecord(MyRecord As tSecurityRecord)
    As Boolean
WriteSecurityRecord = PDBWriteRecord(dbSecurity, VarPtr(MyRecord))
End Function
```

Listing 8.10 *(Continued)*

The only piece of code in Listing 8.11 that is not generated by AppForge is function fnGetFieldByIdx (lines 72 through 76). We have added this function to assist us in search functionality.

```
'-------------------------------------------------------------
'     AppForge PDB Converter auto-generated code module
'
'  Source Database: C:\book2\AppForge Application\PolAlert.mdb
'     Source Table    : Wanted
'
'     Num Records     : 3
'
'     PDB Table Name : Wanted
'           CreatorID : 12sv
'           TypeID    : DATA
'
'     Converted Time : 1/5/2002 6:19:42 AM
'
'-------------------------------------------------------------
Option Explicit
' Use these constants for the CreatorID and TypeID
Public Const Wanted_CreatorID As Long = &H31327376
Public Const Wanted_TypeID As Long = &H44415441
' Use this global to store the database handle
Public dbWanted As Long
' Use this enumeration to get access to the converted database Fields
Public Enum tWantedDatabaseFields
RecordNo_Field = 0
LastNAme1_Field = 1
FirstName_Field = 2
OnTheRun_Field = 3
Chars_Field = 4
Dengerous_Field = 5
End Enum
Public Type tWantedRecord
RecordNo As Long
LastName As String
FirstName As String
OnTheRun As Boolean
Chars As String
Dengerous As Boolean
End Type
Public RecToShow As tWantedRecord
Public loadRec As Boolean
Public Function OpenWantedDatabase() As Boolean
' Open the database
#If APPFORGE Then
```

Listing 8.11 AppForge-generated code for the PolAlert table.

```
dbWanted = PDBOpen(Byfilename, "Wanted", 0, 0, 0, 0, afModeReadWrite)
#Else
dbWanted = PDBOpen(Byfilename, App.Path & "\Wanted", 0, 0, 0, 0,
afModeReadWrite)
#End If

If dbWanted <> 0 Then
'We successfully opened the database
OpenWantedDatabase = True

Else
'We failed to open the database
OpenWantedDatabase = False
#If APPFORGE Then
MsgBox "Could not open database - Wanted", vbExclamation
#Else
MsgBox "Could not open database - " + App.Path + "\Wanted.pdb" +
    vbCrLf + vbCrLf + "Potential causes are:" + vbCrLf + "1. Database
    file does not exist" + vbCrLf + "2. The database path in the
    PDBOpen call is incorrect", vbExclamation
#End If
End If

End Function

Public Sub CloseWantedDatabase()

' Close the database
PDBClose dbWanted
dbWanted = 0

End Sub

Public Function ReadWantedRecord(MyRecord As tWantedRecord) As
    Boolean
ReadWantedRecord = PDBReadRecord(dbWanted, VarPtr(MyRecord))
End Function

Public Function WriteWantedRecord(MyRecord As tWantedRecord) As
    Boolean
WriteWantedRecord = PDBWriteRecord(dbWanted, VarPtr(MyRecord))
End Function

Public Function fnGetFieldByIdx(ByVal lThisIndex As Long) As
String
Dim sThisValue As String   'Temp var for returned field value
PDBGetField dbWanted, lThisIndex, sThisValue     'Use PDBGetField
    to retrieve field value
fnGetFieldByIdx = sThisValue 'Return value with function name
End Function
```

Listing 8.11 *(Continued)*

At this point, we have completed our code and we did initial testing in two stages:

1. Test via the VB IDE to ensure that we do not have any "design flaws."
2. Test on the Palm Emulator to see how the layout of the fields will fit on the screen of the small Palm-based device.

The time has come to send all of the components of our application to the Palm-powered device. This is as simple as selecting the "Compile and validate" and "Deploy to Palm OS" options from the VB IDE. This is correct as long as there are no error messages, but even then we are provided with help.

AppForge adds some interesting features that should help you in deploying your final product to the Palm-based device. For example, we intentionally changed StartUpPosition for the form frmPersOpt to Manual. AppForge detected this setting as potentially conflicting for the final outcome (as you can see from the message in Figure 8.17) of the compilation process. Therefore, it will act as a guide that will point us in the right direction and help us to prevent and solve possible errors.

One more important part of the AppForge application environment is to correctly configure AppForge settings (also accessible from VB IDE). Here you can set details such as dependencies (in this case, the Security and Wanted databases) as shown in Figure 8.18.

Some other details that can be set are Creator ID (as we discussed at the beginning of this chapter), HotSync name, whether to install Booster, and whether to create backup copies of the applications and the databases. These are just some of the settings that you will use more often (see Figure 8.19).

What else is left to be done? Ah, yes, please press the HotSync button on your Palm cradle to sync up your application with the Palm device and start your happy hunt for the fugitives. Since our goal in this chapter was to introduce you to the structure of PDB databases and to enable you to quickly develop and deploy PDB database-driven applications, we did not cover all of the

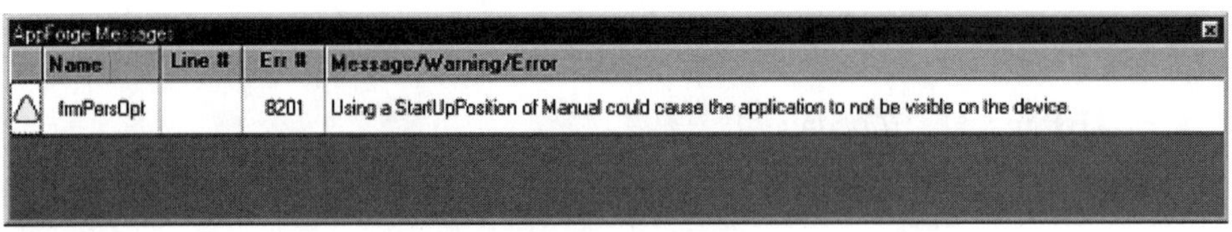

Figure 8.17 AppForge compilation warning message.

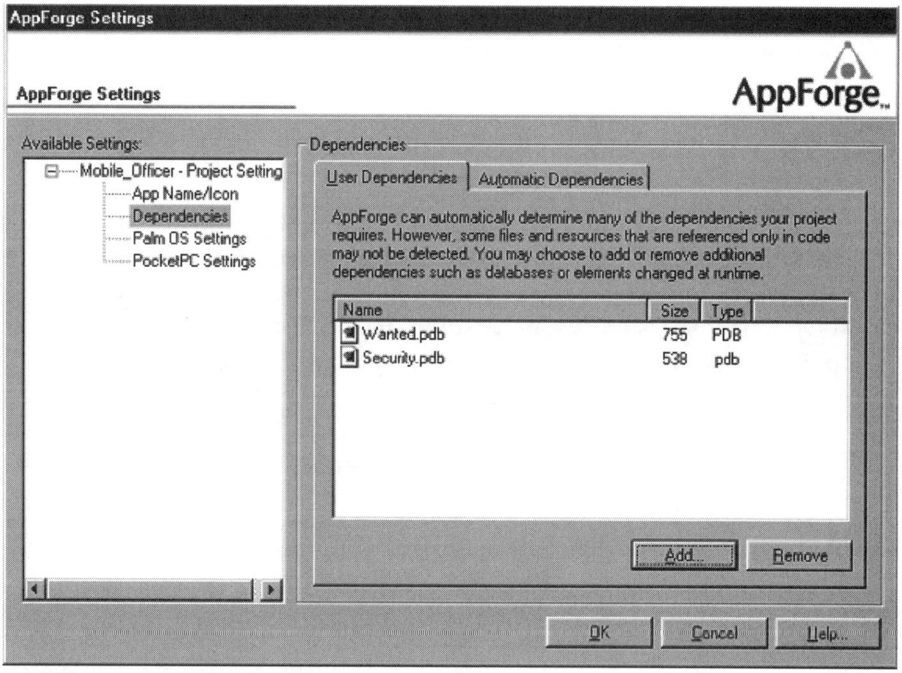

Figure 8.18 AppForge settings 1.

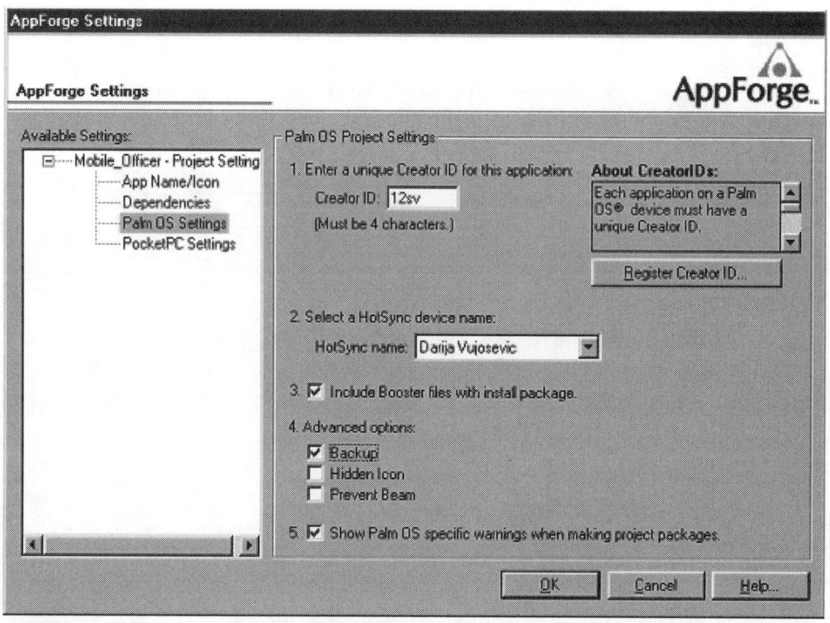

Figure 8.19 AppForge settings.

details regarding the AppForge add-on for VB. For more details about this product, please visit www.appforge.com.

Final Thoughts

As you can see from this chapter, built-in support for the PDB databases makes this type of database an excellent choice for the simple projects. Availability of the tools for the application development that directly "talk" to the PDB databases is one more reason to use PDB-type databases as your starting point for portable application development. However, since PDB databases have limitations (e.g., the maximum number of records is 64K), you should consider using an enterprise-class Relational Database Management System (RDBMS) for the projects that are considered "mission critical."

CHAPTER 9

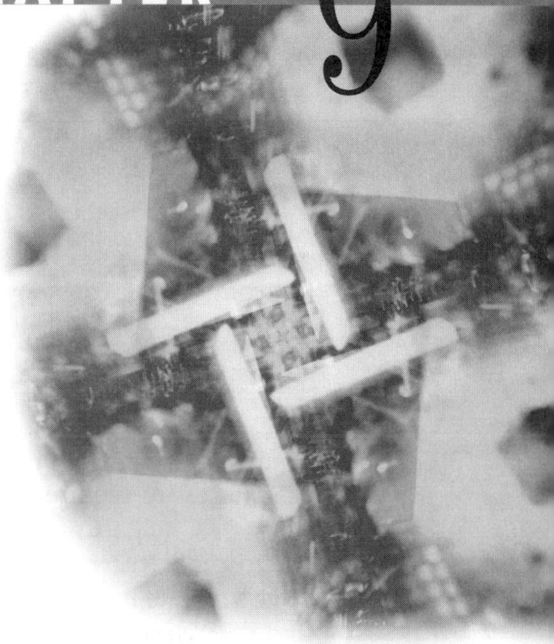

Microsoft

In this chapter, we will introduce part of Microsoft's offering in the arena of mobile databases with special attention to database creation, database replication and synchronization, application development, and features of Microsoft SQL 2000 CE edition in cooperation with ADOCE and eMbedded Visual Basic (eVB). Let's start with the list of the items you will need to establish the infrastructure required for the example provided in this chapter. To clarify, the architecture presented here is a generic one and is in no way specific for the example presented in this chapter.

Each of the software components required is available as evaluation downloads from Microsoft:

- Microsoft SQL Server 2000 Standard Edition (evaluation is available for download from www.microsoft.com/sql/evaluation/trial/2000/default.asp)
- Microsoft SQL Server 2000 CE Edition (evaluation is available for download at www.microsoft.com/sql/evaluation/trial/CE/download.asp)
- Microsoft Windows 2000 Server
- Microsoft Windows 2000 Server Service Pack 2 (available for download from www.microsoft.com/windows2000/downloads/servicepacks/sp2/default.asp)
- Microsoft SQL Server 2000 Service Pack 2 (available for download from www.microsoft.com/sql/downloads/2000/sp2.asp)

- Microsoft SQL Server 2000 CE Edition Service Pack 1 (available for download from www.microsoft.com/sql/downloads/CE/v11SP1.asp)
- Update pack for Microsoft SQL Server 2000
- Microsoft Embedded Visual Tools version 3.0 (available for download at www.microsoft.com/mobile/downloads/emvt30.asp)
- Microsoft Pocket PC 2002 SDK (available for download at www.microsoft.com/mobile/developer/downloads/ppcsdk2002.asp)
- Active Sync 3.5 (minimum version is 3.1—delivered with Compaq iPAQ Pocket PC)

Sounds like a lot of components, but in fact, all of them are nicely integrated. For hardware requirements, we have two options. One is to have a single Pocket PC (e.g., Compaq iPAQ 3800 series) and one PC that will serve as Microsoft SQL server, IIS server, and ActiveSync host. Another option is to have separate PCs for each of the major components (docking PC, SQL server, and IIS Web server). For reasons of simplicity we will select the first option for this chapter. Keep in mind that your PC must be able to run Windows 2000 and SQL 2000. The minimum PC requirements are:

- 128MB of RAM (256MB recommended)
- Pentium III CPU "ticking" at 500 MHz minimum
- 200–300MB of free hard disk space
- Available USB or RS-232 (usually known as Com port)
- Windows CE powered device (e.g., Compaq iPAQ 3800)
- An active network connection

SQL Server 2000 CE Edition

More and more, portable devices are becoming delivery channels for enterprise-level applications. These small devices have increased capabilities, with memory, storage, and display features closer to those of their "bigger siblings," such as desktop PCs and laptops. However, physical device limitations are still presenting challenges for the developers. Additional challenges arise from the fact that connectivity to the base unit, LAN, or Internet can be intermittent, lower capacity, or nonexistent.

The SQL Server CE edition addresses all of the preceding issues. Unlike built-in CDB databases, the limitations of SQL Server CE are 2GB database size and

1GB for BLOB objects. This small footprint occupies only 800KB to 1.3MB of portable device memory—this is based on portable platform type. Even in this small footprint, SQL Server CE supports the following features:

- Vertically compatible SQL grammar with SQL Server 2000.
- The following data types are supported:
  - TINYINT, SMALLINT, INTEGER, BIGINT
  - BIT, VARBINARY, BINARY, IMAGE
  - REAL, NUMERIC, FLOAT, DOUBLE PRECISION
  - UNICODE NATIONAL CHARACTER, NATIONAL CHARACTER VARYING, NTEXT
  - MONEY, DATETIME, UNIQUEIDENTIFIER
- Up to 32 indexes per table.
- Support for NULL values and nested transactions.
- 128-bit file-level encryption for the database file (must be also password protected).
- DDL commands: Create databases, alter tables, referential integrity, and default values.
- DML statements: INSERT, UPDATE, and DELETE.
- Date and time functions such as DATEADD, DATEDIFF, DATENAME, DATEPART, and GETDATE.
- Support AVG, COUNT, MAX, MIN and SUM aggregate functions, INNER/ OUTER JOIN, nested SELECT statements, GROUP BY/HAVING
- Scrollable and forward-only cursors

In order to interoperate with the other Microsoft SQL products, the CE edition of SQL Server supports remote data access (RDA), merge replication, and encryption of data transfer. This is the place where IIS 5.0 comes into play, since all of the preceding methods are enabled via HTTP (we will see more about this during the install phase of SQL Server CE).

It is also important to note that Microsoft introduced OLEDBCE and ADOCE to enable ADO "savvy" developers to adopt development for CE-based devices. Each of these products is somewhat limited in the functionality provided. For example, compared to the ADO model, ADOCE provides the functionality listed in Table 9.1.

Since communication with the Enterprise class databases is performed over secured HTTP protocol, implementation of the CE solutions is straightforward.

Table 9.1 ADOCE Supported and Unsupported Objects

SUPPORTED FEATURE	UNSUPPORTED
Recordset	Connection
Field	Command
Fields	Error
	Property

This is because a majority of the firewalls are already configured to allow traffic to ports 80 (standard HTTP) or 443 (SSL encrypted HTTP).

An additional feature that is often handy is the so-called SQL Server CE Relay. This means that while you are connected to your PC via serial or USB ports, you can use a pass-through connection to the backend database. A pass-through connection allows you to use desktop (or laptop) Internet connection from the Pocket PC. Before pass-through, you had to use either a modem or separate Ethernet adapter to obtain Internet connection. Don't forget—SQL Server CE databases are not case sensitive. You cannot perform replication between an SQL Server CE and a case-sensitive SQL Server database. This also limits your options for the selected database dictionary if you set up replication with Windows SQL CE subscribers.

Installing Windows SQL 2000 CE Edition

Before you start installation of the required products, you must define the environment you will use for the configuration. Here we will present you with two configurations that are most commonly used. The first is the maximum configuration that will enable you to simulate a near real-life environment (Figure 9.1), and the second is the minimum that most of us can put together to use the example created in this chapter (see Figure 9.2).

For the maximum (or even better "complete configuration"), you will need two servers (SQL and IIS server), a workstation, and a Windows CE-powered device connected to the workstation. We know, it is too much for the average person to put together. Let's see what is needed to install the required components.

You must install the following software on your development system before you can start developing SQL Server CE-based applications.

1. Development workstation installation.
   - Microsoft ActiveSync 3.1 or later (the latest version is delivered with your Windows CE-based device).

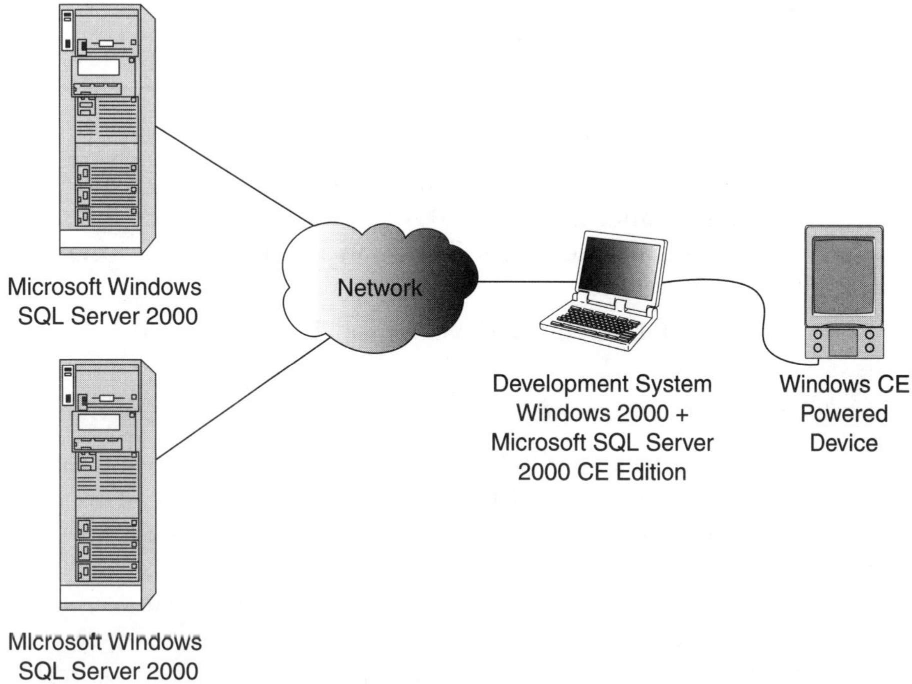

Figure 9.1 Production system configuration.

At the time this chapter was written, ActiveSync 3.5 was the only version available. If (and this is usually true for small development environments) you are using ActiveSync with USB, IR, or serial communication between a Windows CE device and your desktop PC (ActiveSync will be used in our example as well), you must install the SQL Server CE Relay component. SQL Server CE Relay is installed automatically on your development system at the same time you install the development tools in the directory {install path}\redist\relay. The executable name is sscerelay.exe. By executing the following command line, the relay component will be installed in such a way that it will automatically start every time an ActiveSync connection is made.

```
sscerelay /clientport 81 /servername {server name} /serverport 80 /register
```

Let's analyze the meanings of the different switches in the command line for a moment:

- The mobile device listens on port 81.
- The desktop machine listens on port 80.

   - The proxy server is named {server name}.
   - The /register option enables Relay to be run every time an ActiveSync connect is made.
- Microsoft eMbedded Visual Tools 3.0.
- Microsoft Windows CE Platform Builder Toolkit 3.0. It is also important to install the Platform Builder 3.0 Add-On Pack. You need the Add-On Pack in order to have the ADO-CE (version 3.1) components installed properly for Platform Builder.
- Microsoft SQL Server CE Development Tools.

2. IIS System installation.

   In order to use either RDA or replication, you must install components of Windows SQL Server CE components on the IIS system. This can be achieved by executing installation of Microsoft SQL Server CE Development Tools and selecting the Server Tools checkbox on the setup screen. Another way of doing this is to install Server Tools on to your development computer and copy the content of the {install path}\redist\server directory from your development workstation to your IIS system. After the copying is completed, execute the server.exe program (on the IIS system) and simply follow the setup screens. Don't forget to install the same release (version) of the development and IIS server components.
3. SQL Server 2000 installation.

   If you have applied all of the patches and upgrades noted at the beginning of this chapter, there are no additional components or scripts that you have to run.

4. Installation of Windows SQL CE components onto your CE device.

   The easiest way to install all of the required components is to use eMbedded Visual Basic (eVB) development tools to download your new SQL Server CE application and all of the required components directly into a Windows CE device using an ActiveSync connection

It is also possible to use just your development workstation for configuration (in combination with ActiveSync and a Windows CE-powered device) as shown in Figure 9.2.

In this scenario, repeat steps 1 through 4 from the previous installation on the development workstation. Your only choice of operating system in this case will be Windows 2000 Server. This configuration can satisfy small development projects that are designed mostly as proof of concept.

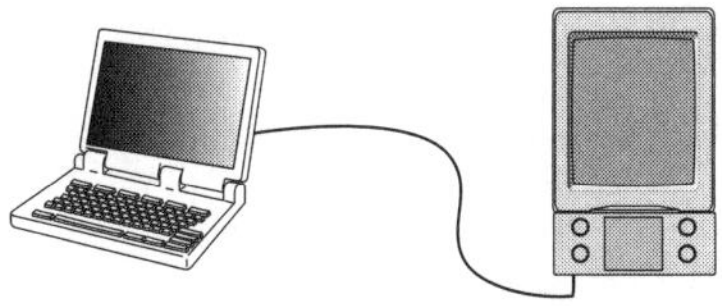

Figure 9.2 Development system configuration.

File Locations

Tables 9.2 through 9.5 will help you locate the files that will be installed during the setup of the required components. Please note that if you are installing all of the components on the same PC (apart from Windows CE device components), the file location will be under the same "root" file system on the development PC.

Development System

The default location for installation is in the "\Program Files\Microsoft SQL Server CE" folder (Table 9.2).

Internet Information Services System

The default location for the installation is in the "\Program Files\Microsoft SQL Server CE" folder. In addition, you should have at least Microsoft Data Access Component version 2.6 or later installed (Table 9.3).

ActiveSync System

The default location for the installation is in the "\Program Files\Microsoft SQL Server CE" folder (Table 9.4).

Windows CE Device

File location on the Windows CE device (Table 9.5).

Table 9.2 Windows SQL 2000 CE Edition File Location on a Development System

LOCATION	CONTENTS
\Device	Multiple subfolders that contain five DLLs (Ssce10, Ssceca10, Sscecw10, Ssceinet, and Sscesock). There is a copy of the DLLs for use with each of the processors supported by SQL Server CE. Each subfolder is not the same in its content.
\Encryption	Multiple subfolders used to implement security features of SQL Server CE. There is a copy of the DLLs (Rsaenh.dll) for each processor and Windows CE operating system version supported by SQL Server CE.
\Inc	Include files for use with Microsoft eVB and Microsoft eMbedded Visual C++.
\Lib	Multiple subfolders that contain the library Ca_mergex.lib for the SQL Server CE ActiveX objects. There is a separate folder for use with each processor supported by SQL Server CE.
\Relay	The SQL Server CE Relay executable (SSCERelay.exe).
\Redist	Multiple subfolders that contain components that you can include, if required, in your deployment package when you have finished your application.

Table 9.3 Windows SQL 2000 CE Edition File Location on an IIS System

LOCATION	CONTENTS
\Server	SQL Server CE Server Agent DLL (sscesa10.dll)
\Server	SQL Server CE replication DLL (sscerp10.dll)
\Program Files\Microsoft SQL Server\80\com	SQL Server CE replication components

Table 9.4 Windows SQL 2000 CE Edition File Location on an ActiveSync System

LOCATION	CONTENTS
\Relay	SQL Server CE Relay executable (SSCERelay.exe)

Table 9.5 Windows SQL 2000 CE Edition File Location on Windows CE

LOCATION	CONTENTS
\Windows	Ssce10.dll, Ssceca10.dll, Sscecw10.dll, Rsaenh.dll, ADOCE, OLE DB files, and Ssceinet.dll and Sscesock.dll, required by RDA or replication process.

Replication

The transport mechanism to perform synchronization for SQL Server CE is HTTP. This connection is established from SQL Server CE to Microsoft SQL Server 2000 through Microsoft Internet Information Services. This type of connectivity is well suited for the modern infrastructure, and connections can be established to the SQL servers hidden behind proxy servers and firewalls. Replication can be established via LANs, WANs, and PC network connection (remember Relay functionality explained earlier). It is important to note that encryption, compression, authentication, and authorization are based on existing IIS features. Additional functionality can be achieved by implementing Microsoft Message Queue services to safeguard against intermittent or poor-quality connections.

In this chapter, we will present creation and usage of merge replication in SQL Server CE. This type of replication is based on Microsoft SQL Server 2000 merge replication and is ideally suited to portable devices because it enables data to be updated independently on the portable device and the server. Data can be synchronized at any point (once the device is connected to the Microsoft SQL Server). Ideal applications for the merge replication would be data collection and upload to the central database (this is the type we will use for the example), or read-only replication where, for example, an "on the road" salesperson can have an up-to-date price list of the products that he or she is selling.

Replication can be set to use horizontal (row filtering) and vertical (column filtering). This way, we can provide specific subsets of the data for the appropriate clients or groups of clients.

MERGE REPLICATION STEPS

1. The publication must be defined on the SQL server. The publication is nothing more than a set of articles defined by the database administrator. An article is a table that is defined for replication.
2. The blank database (extension .sdf), which will hold replication data, must be created (or delivered as part of the distribution package) on the Windows CE device.
3. As soon as the publication is defined and created by the database administrator (DBA), a Windows CE application can subscribe to the publication by executing the Replication Object. The published data is then downloaded to the SQL Server database on the Windows CE device.
4. The data in the Windows SQL 2000 CE database can then be updated, changed, or deleted by applications running on the Windows CE device.

5. Our application will trigger synchronization by executing a Replication Object. Synchronization can be created in the application as an automated process (e.g., as soon as the device is connected to the server), or it can be initiated by the end user (select File->Synchronize option in our database). When synchronization is performed, the updates made to the replica are sent to the publisher where they are merged into the publisher's database. Similarly, changes made by the publisher since the initial download or most recent merge are sent to the Windows CE device where they are merged into the database.

Sample Application—Surveys

A couple of weeks ago we were stopped by a young woman who was doing a survey for a transportation company. For each person that she interviewed, she would write the information on a new blank sheet of paper. Once she returned to the office, the data would then be entered into the database. That was a waste. After that, though, it occurred to us that showing how to create an application based on Windows SQL CE and data replication would be a great example for this chapter. With an application like this one, data will be entered only once—there would be no messy paperwork. To maintain simplicity, we will assume that you will be running all of the required components on the same computer.

Before we start database design and development, we have to create an application flow diagram that will help us design, code, and gain a general overview of the sample application idea. Application flow for the survey application is presented in Figure 9.3.

From the application flow, you can see that we will create the application with a relatively low level of complexity. Figure 9.3 also represents the goal of this small application, which is to collect information "on the streets" and propagate the collected data to one central database. Once we understand what our application is all about we can build the "Survey" database and application.

For this example, we have three major parts of the application to complete:

1. Create the Survey database.
2. Set up replication.
3. Create the Survey application.

In the real-life cycle of portable database and application development, teams of specialists will perform the work. A DBA, application developers, and sys-

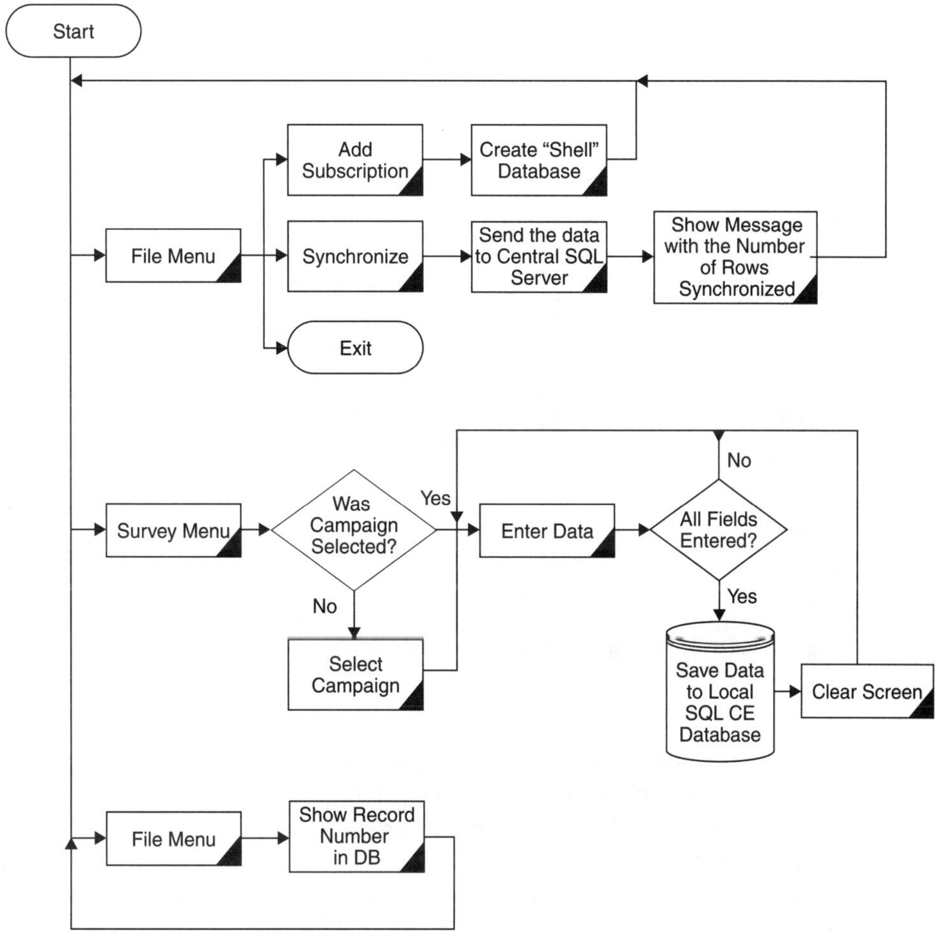

Figure 9.3 Application flow diagram.

tem integrators are only minimum requirements for mobile application design and development.

In this chapter, we will guide you through the steps required to build the Survey application. The steps will involve knowledge and skill sets from different fields. Let's get to work.

Create the Survey Database

As you can see in Figure 9.4, there will be only two tables in our database. Table tblSurveys will hold information about all of the surveys that can be performed at the moment, and table tblSurveyResults will hold collected data.

tblSurveyResults

Column Name	Data Type	Length	Allow Nulls
ID	int	4	
Question1	char	10	✓
Question2	char	10	✓
Question3	char	10	✓
Question4	char	10	✓
Question5	char	10	✓
SurveyID	char	10	
Upload	char	1	✓
rowguid	uniqueidentifie	16	

tblSurveys

Column Name	Data Type	Length	Allow Nulls
SurveyID	int	4	
SurveyTitle	char	20	
rowguid	uniqueidentifie	16	

Figure 9.4 Survey application table structure.

During the table design, you must plan in advance, especially if you are going to use replication, as in our case. For example, we will be using horizontal filtering (row-level filtering) to make sure that rows of collected data will be synchronized to the central database only once. For that reason, we have added an Upload field. The purpose of it will be explained later in this chapter.

To simplify creation of the database and required tables, please use the script in Listing 9.1.

```
/****** Object:  Database Survey
   Script Date: 3/25/2002 12:16:41 AM ******/
IF EXISTS (SELECT name FROM master.dbo.sysdatabases WHERE name =
N'Survey')
    DROP DATABASE [Survey]
GO

CREATE DATABASE [Survey]  ON (NAME = N'Survey_Data', FILENAME =
   N'C:\Program Files\Microsoft SQL Server\MSSQL\data\
   Survey_Data.MDF' , SIZE = 2, FILEGROWTH = 10%) LOG ON (NAME =
   N'Survey_Log', FILENAME = N'C:\Program Files\Microsoft SQL
   Server\MSSQL\data\Survey_Log.LDF' , SIZE = 1, FILEGROWTH = 10%)
```

Listing 9.1 SQL script to create required database and tables.

```
COLLATE SQL_Latin1_General_CP1_CI_AS
GO

exec sp_dboption N'Survey', N'autoclose', N'false'
GO
exec sp_dboption N'Survey', N'bulkcopy', N'false'
GO
exec sp_dboption N'Survey', N'trunc. log', N'false'
GO
exec sp_dboption N'Survey', N'torn page detection', N'true'
GO
exec sp_dboption N'Survey', N'read only', N'false'
GO
exec sp_dboption N'Survey', N'dbo use', N'false'
GO
exec sp_dboption N'Survey', N'single', N'false'
GO
exec sp_dboption N'Survey', N'autoshrink', N'false'
GO
exec sp_dboption N'Survey', N'ANSI null default', N'false'
GO
exec sp_dboption N'Survey', N'recursive triggers', N'false'
GO
exec sp_dboption N'Survey', N'ANSI nulls', N'false'
GO
exec sp_dboption N'Survey', N'concat null yields null', N'false'
GO
exec sp_dboption N'Survey', N'cursor close on commit', N'false'
GO
exec sp_dboption N'Survey', N'default to local cursor', N'false'
GO
exec sp_dboption N'Survey', N'quoted identifier', N'false'
GO
exec sp_dboption N'Survey', N'ANSI warnings', N'false'
GO
exec sp_dboption N'Survey', N'auto create statistics', N'true'
GO
exec sp_dboption N'Survey', N'auto update statistics', N'true'
GO
use [Survey]
GO

/****** Object:  Table [dbo].[tblSurveyResults]
    Script Date: 3/25/2002 12:16:41 AM ******/
if exists (select * from dbo.sysobjects where id =
    object_id(N'[dbo].[tblSurveyResults]') and OBJECTPROPERTY(id,
    N'IsUserTable') = 1)
drop table [dbo].[tblSurveyResults]
GO
/ • • • • • • • Object:  Table [dbo].[tblSurveys]
    Script Date: 3/25/2002 12:16:41 AM • • • • • •/
if exists (select     • from dbo.sysobjects where id =
    object_id(N'[dbo].[tblSurveys]') and OBJECTPROPERTY
    (id, N'IsUserTable') = 1)
```

Listing 9.1 (Continued)

```
drop table [dbo].[tblSurveys]
GO

/****** Object:  Table [dbo].[tblSurveyResults]    Script Date:
3/25/2002 12:16:53 AM ******/
CREATE TABLE [dbo].[tblSurveyResults] (
    [ID] [int] IDENTITY (1, 1) NOT NULL ,
    [Question1] [char] (10) COLLATE SQL_Latin1_General_CP1_CI_AS NULL ,
    [Question2] [char] (10) COLLATE SQL_Latin1_General_CP1_CI_AS NULL ,
    [Question3] [char] (10) COLLATE SQL_Latin1_General_CP1_CI_AS NULL ,
    [Question4] [char] (10) COLLATE SQL_Latin1_General_CP1_CI_AS NULL ,
    [Question5] [char] (10) COLLATE SQL_Latin1_General_CP1_CI_AS NULL ,
    [SurveyID] [char] (10) COLLATE SQL_Latin1_General_CP1_CI_AS NOT NULL ,
    [Upload] [char] (1) COLLATE SQL_Latin1_General_CP1_CI_AS NULL ,
    [rowguid]  uniqueidentifier ROWGUIDCOL  NOT NULL
) ON [PRIMARY]
GO

/****** Object:  Table [dbo].[tblSurveys]    Script Date: 3/25/2002
12:16:55 AM ******/
CREATE TABLE [dbo].[tblSurveys] (
    [SurveyID] [int] IDENTITY (1, 1) NOT NULL ,
    [SurveyTitle] [char] (20) COLLATE SQL_Latin1_General_CP1_CI_AS NOT NULL ,
    [rowguid]  uniqueidentifier ROWGUIDCOL  NOT NULL
) ON [PRIMARY]
GO
```

Listing 9.1 SQL script to create required database and tables. *(Continued)*

After you run the script in Listing 9.1, the Survey database should appear in the list of available databases (in Enterprise Manager). Grant your local IUSR_{your computer name} user SELECT, DELETE, INSERT, and UPDATE rights for the Survey database for all user tables.

NOTE

The IUSR_{your computer name} user has been created by the IIS 5 install process and is used for Anonymous access to the IIS and related resources.

In the real production environment you will most likely limit access to the databases. For example, in our case only SELECT rights are required for tblSurveys, since the changes to this table will be created on the Publisher side.

Replication Setup

Microsoft really did a great job in providing a well-designed GUI interface for the creation of different types of replication (part of the MS SQL 2000 Enterprise Manager). Before we start using this wizard, let's create a folder in the

file system called Cerepl and grant our anonymous IUSR all the rights for this directory. To help you prepare your file system for the sample application, follow the next step-by-step process, which will help you configure correct rights for the database replication setup.

1. From Windows Explorer, create a new NTFS file system directory. Name this directory "CEREPL". This directory will contain the snapshot folder.
2. Right-click the new directory, select Properties, and then select the Sharing property tab. Click "Share this folder" and as "Share name," enter the same name we used for the directory name, "CEREPL".
3. Click the Permissions button to display the Share Permissions page. Click the Add button to grant permissions (outlined in Table 9.6) to the share.
4. Close all of the opened pages by simply clicking the OK button.

Setting up database publication is a straightforward task when you are using the Create Publication wizard. For an alternative method to creating the required replication, we have prepared a SQL script (see Appendix B, "Microsoft Publication Wizard Script") that will create the same publication as the step-by-step process we will execute in the next segment of this chapter. If you take a moment to look at the script in Appendix B, the level of SQL language knowledge required to create publications with the SQL script will become obvious. Now let's start the Create Publication wizard by selecting Tools->Wizards in Windows SQL 2000 Server Enterprise Manager utility. The first screen is the database selection screen (see Figure 9.5) where you will select the database to use for the new publication.

Table 9.6 Required Replication User Rights

USER	REQUIRED PERMISSIONS
SQL Server Service account and SQL Server Agent Service account.	Full Control
For IIS Anonymous Authentication, grant IUSR_*computername* or the configured IIS anonymous user account read permission. In this case, IUSR_*computername* is the local account on the IIS system, and IIS and SQL servers are physically located on the same PC (in most "proof of concept" environments this would be the case). In case the IIS and SQL systems are located on two separate PCs, IUSR_*computername* account must be created as the domain user account, consequently, this domain account will have to be granted read access to "CEREPL" share.	Read

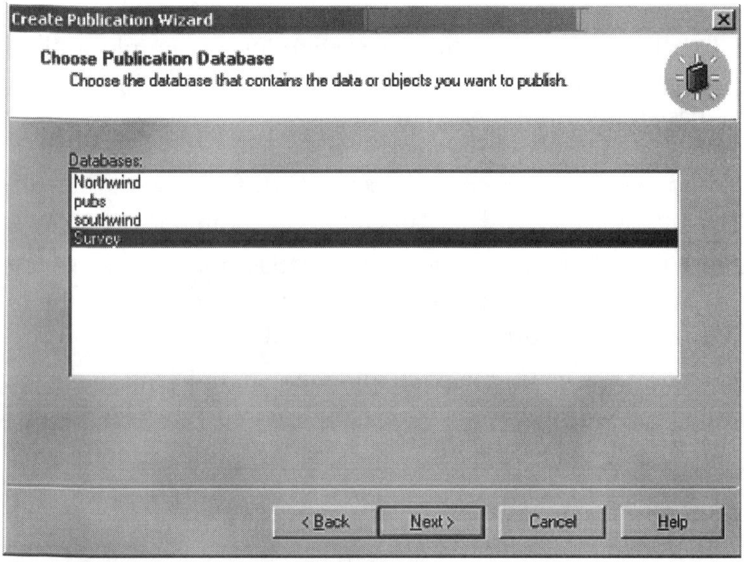

Figure 9.5 Select publication database.

Select the Survey database, and click Next.

You will then have to select the type of Publication you want to create (see Figure 9.6). You have a choice of a Snapshot, Transactional, or Merge publication. Since SQL Server CE only supports merge publications, our choice for this

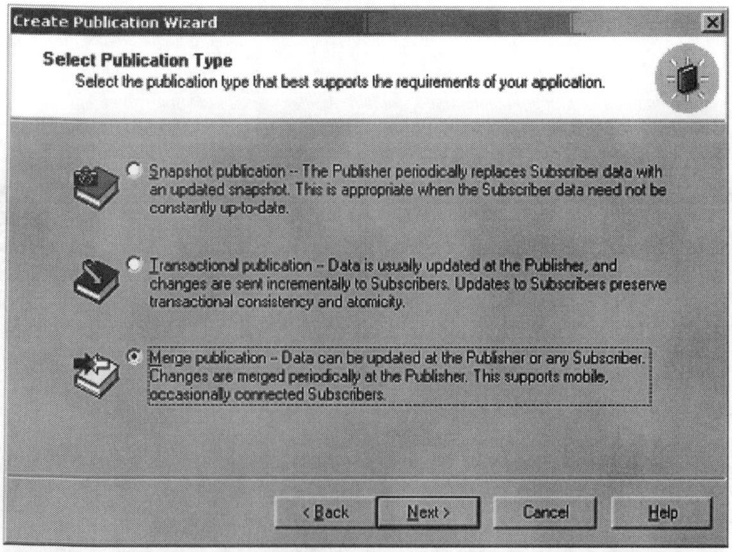

Figure 9.6 Select publication type.

screen has already been predetermined. The other types listed are for the cases when replication will be performed between Windows SQL servers.

Merge replication enables data to be updated on the portable device and the server. The data on the device and on the server can be synchronized when the device is connected to the server (by way of the LAN/WAN Internet connection or through the relay application installed on your development system).

Some of the wired and wireless connections supported by Windows SQL CE remote data access include:

- Wireless LAN protocols (e.g., OpenAir and IEEE 802.11)
- Wireless WAN protocols (e.g., CDMA, GSM, TDMA, CDPD, RAM, AMPS, and ARDIS)
- SQL Server CE Relay-based connections (e.g., USB, serial, and infrared)

Merge replication also allows the developer or database administrator to create vertical (column level) filtering and/or horizontal (row level) filtering. This feature is important in applications when we would like to upload a specific subset of the database to the device (e.g., in cases where we might be using delivery type applications or when we would like to provide specific products to specific salespeople).

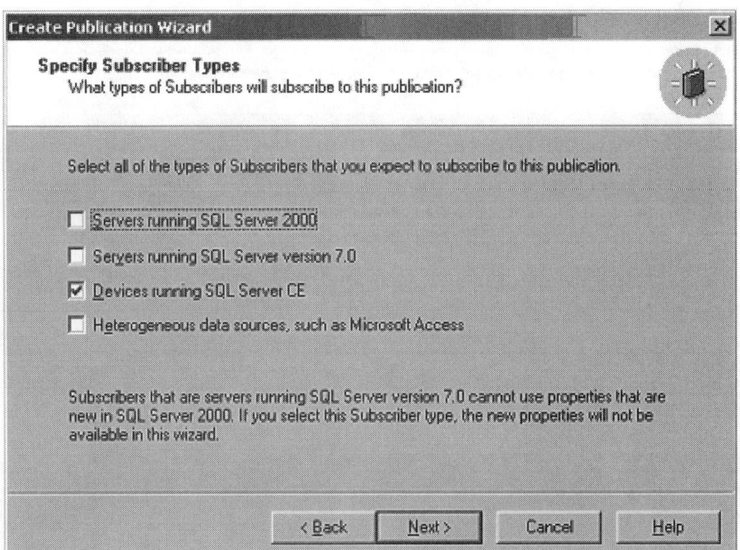

Figure 9.7 Select subscriber type.

After selecting the replication type and clicking Next, you will be asked to select the type of subscribers you will be serving (see Figure 9.7). In our case, it will only be devices running Windows SQL CE. Some applications will use multiple types of subscribers, such as with Enterprise Resource Planning (ERP) systems, where multiple delivery channels are available for data collection and replication.

Select the "Devices running SQL Server CE" option, and click Next. The screen shown in Figure 9.8 should appear.

Here you will have to select articles (in our case, two tables), but for applications that are more complex, you can see views and stored procedures in the list of available objects. Although SQL Server publication can contain other database objects, such as stored procedures, views, user-defined functions, and triggers, the SQL Server CE replication engine ignores these objects and does not include them in your SQL Server CE subscription. For our sample application, select the two tables (tblSurveys and tblSurveyResults) that are required for our example, and click Next. The resulting screen is mostly informational and tells us a bit more about the requirements (during the database design phase) for unique identifiers for each row. The Create Publication wizard will detect that you have missed this requirement and will create unique identifiers for you. We could have done this before starting the wizard, but we intentionally decided to "forget" this step and allow the wizard to correct this mistake for us.

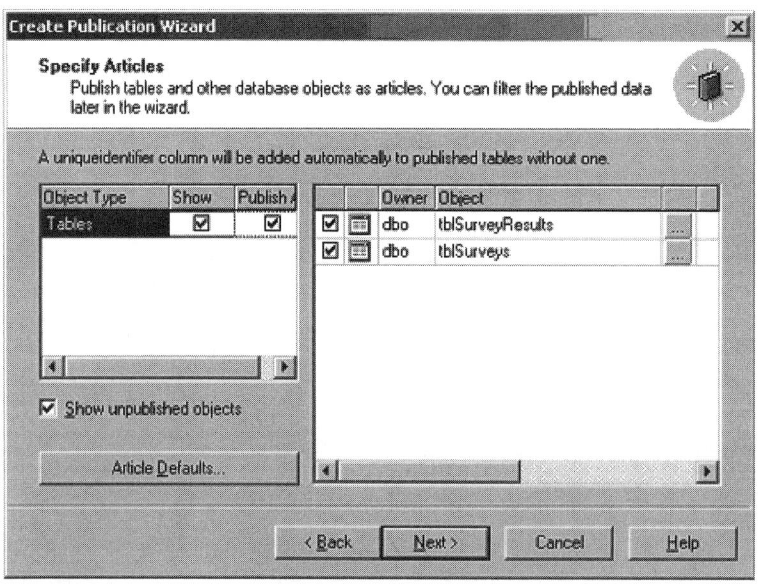

Figure 9.8 Select publication articles (tables).

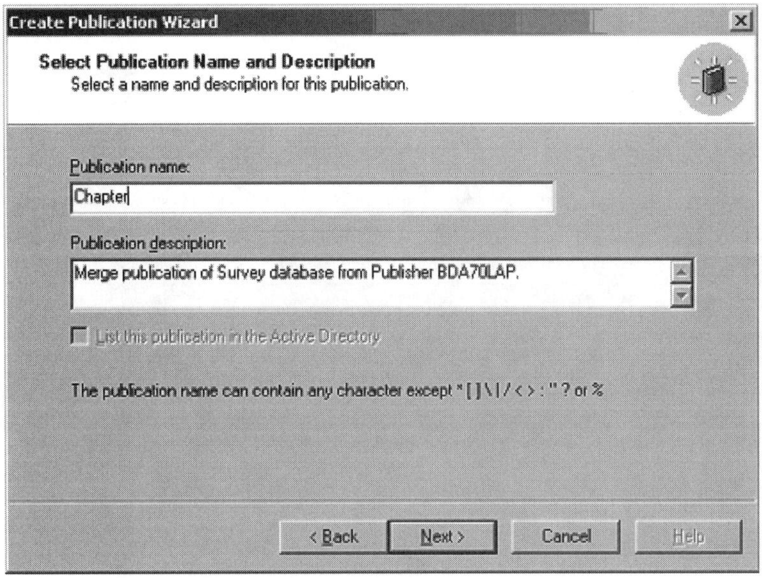

Figure 9.9 Assign publication name.

Accept the suggested changes to the table structure by clicking Next.

Our next task will be to assign the publication name (see Figure 9.9). It is important to remember the name you will assign, since this will be used later during the application development phase.

For consistency with scripts provided for this chapter, use the name "Chapter".

At this point, you can elect to create the publication without any additional options, or you can click on the radio button that will open data filtering, anonymous subscriptions, synchronization performance, and schedule for synchronization execution (which we will do in our example, shown in Figure 9.10). When you click Next, you will go to a new screen (see Figure 9.11), which enables you to select the type of filtering (vertical or horizontal) required.

Because we want to filter only rows of data in our example and we want to ensure that records added to the tblSurveyResults table will remain in Publisher database only, you will select only the "Horizontally …." checkbox. By selecting horizontal filtering, we are telling the database engine to provide us with all of the columns and to perform row filtering based on the WHERE statement. Now, remember the Upload field in tblSurveyResults table field structure from Figure 9.4? This field will be used in our sample application to control those rows that will remain in the central SQL database and those rows propagated to the Windows SQL CE database on the portable device. Obviously, we would prefer to

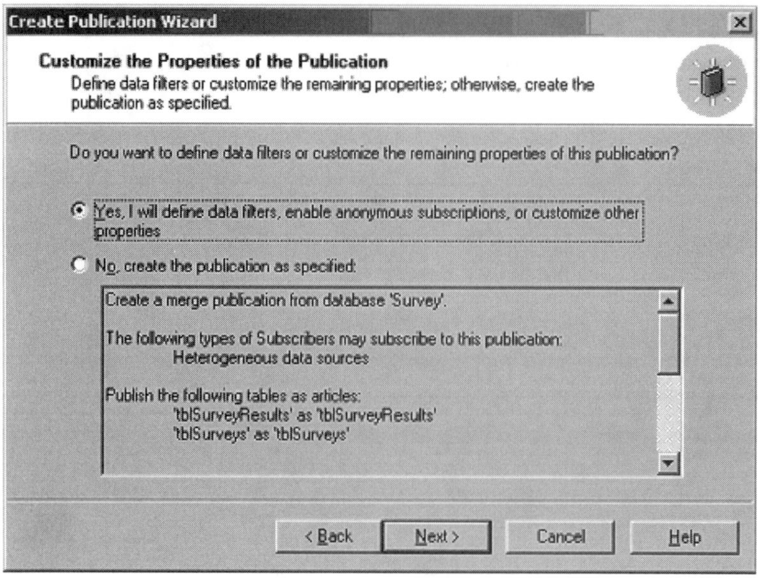

Figure 9.10 Select additional properties.

transfer all of the data to the central database and leave nothing on our portable device, thus preparing us for a new day of collecting surveys.

Clicking Next will bring you to a new screen (see Figure 9.12) that will allow you to select the type of filtering performed—dynamic or static. Static filtering

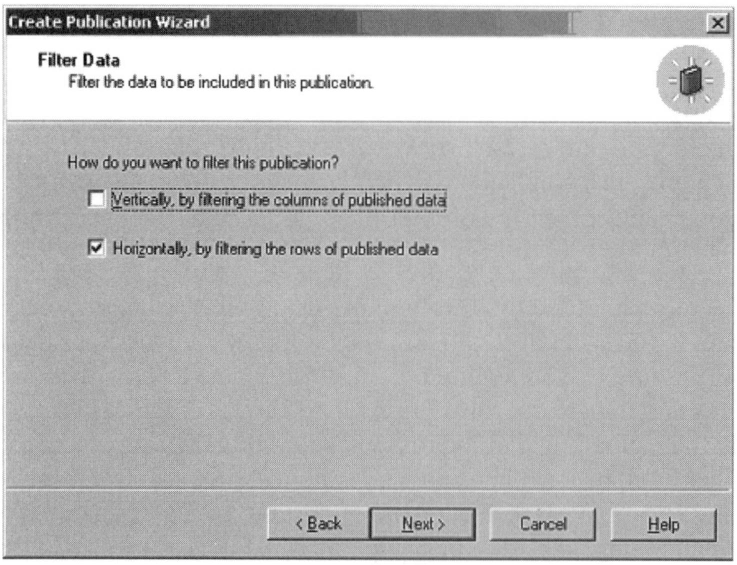

Figure 9.11 Publication wizard filter selection.

(unlike dynamic) provides each subscription device with the same set of data. This is okay for the small number of subscribers and small data sets that have to be synchronized, but if you have a large population of subscribers you should consider dynamic filtering because data synchronization sets are tailored for each subscriber separately. If you have previous experiences with publication subscription methods, you will remember that Windows SQL servers use the "Validate Subscriber Information" option in their replication process. SQL Server CE, on the other hand, uses HostName property to validate subscriber information.

In our application, we will use static filters. What would be a good example if you want to create dynamic filters? One example would be to prepare separate dynamic snapshots for the large population of salespeople that cover different products or different areas. However, let's continue with our application. When you click Next, you will need to specify the SQL statement that will be used to define those rows of data to be propagated to the Windows SQL CE database. You will be asked to complete the WHERE clause in the following SELECT statement:

```
SELECT<published_columns> FROM [dbo].[tblSurveyResults]
   WHERE tblSurveyResults.Upload = 'Y'
```

The most important part of the entire SQL statement is the WHERE clause. This clause will enable us (as you will see during the application development phase) to merge new survey records to Windows SQL database while effectively removing the same records from Windows SQL CE database on your

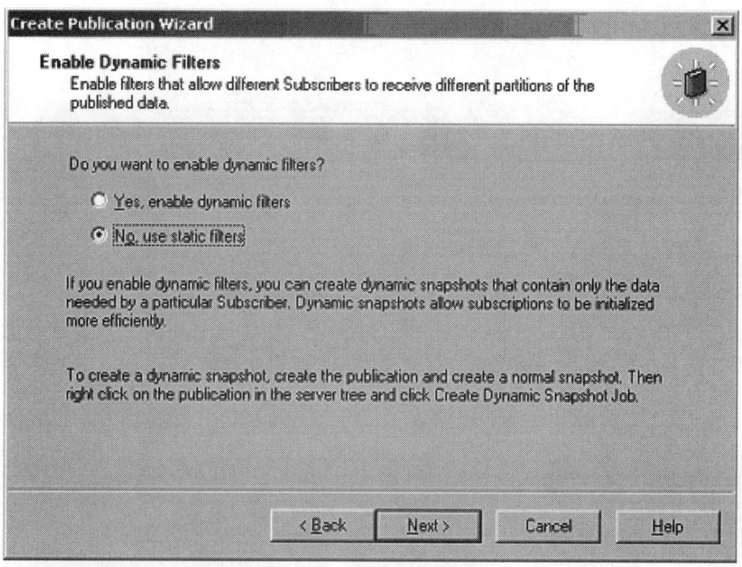

Figure 9.12 Select static or dynamic filters.

portable device. Once you have evaluated the statement and clicked OK, the wizard will begin generating filters for the publication. Once the process is complete, the screen shown in Figure 9.13 will present you with an overview of what has been completed up to this point. If you would like to change any of the options before creating the final publication, this is the place to do so. For the purpose of our sample, we will use this screen to review what we have accomplished so far.

Click OK.

You will be asked if you want to allow anonymous subscriptions to this publication. If you were running anything other than Windows SQL Server CE, you could choose to allow anonymous or named subscriptions. However, because Windows SQL Server CE supports only anonymous subscriptions, that is the only option available.

As we proceed through the wizard, we will be given the option to change/set the time for the execution of the Snapshot Agent.

The reason for the Snapshot Agent is to immediately create a database snapshot file that can be downloaded to your Windows CE-based device. The main pur-

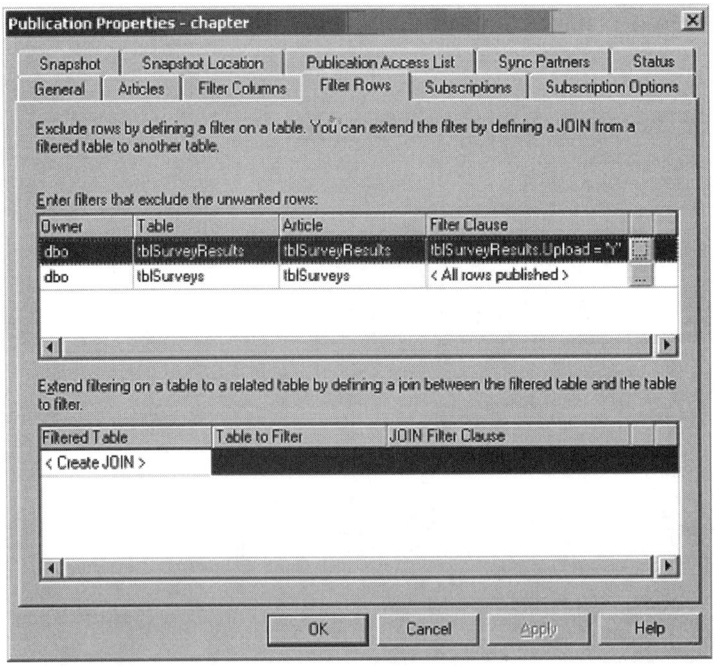

Figure 9.13 Review Publication wizard settings.

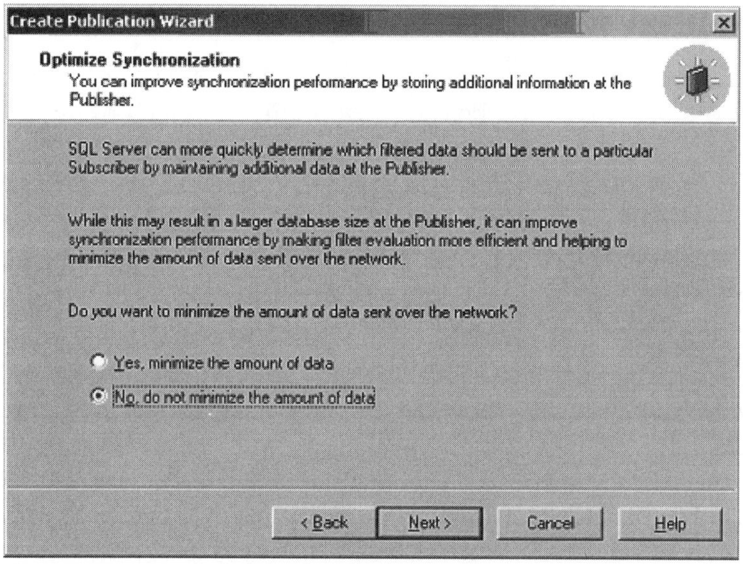

Figure 9.14 Optimize Synchronization options.

pose of the snapshot file is for the initial creation of a database on a portable device and to create a "baseline" that will be used for any future synchronization.

We are almost done creating our publication. When you are satisfied with the schedule you set for the snapshot agent, the wizard will give you an opportunity to review the settings used in this exercise. Clicking the Next button again will bring you to the last option screen, as shown in Figure 9.14.

This screen is important in situations when large amounts of data need to be synchronized because Publisher can determine in advance which data should go to which subscriber, thus saving bandwidth and time required for synchronization. This could be the case when we synchronize with a large data warehouse or something similar, but in this sample application we do not need the space.

Click Next and the SQL Server Enterprise manager will run through the task of creating our publication from the database that we set up. That's the end of the story for creating the sample publication in this chapter. In Appendix B, we have provided the SQL script (generated from the Windows SQL Enterprise Manager application), which should help you to understand the level of skill required to accomplish the same task by writing a publication script from scratch in pure SQL. One of the great achievements in the last couple of years is the advancement in the field of database wizards, which make the lives of less experienced DBAs much easier.

Creating the Survey Application

For those of you who have previously used the Visual Basic development environment, many parts of the code used in this chapter's example application will look familiar. eVB supports many of the "big brother" (VB) keywords. For more details on this topic, refer to Chapter 7, "Mobile Application Development Tools," where we write about development tools, including eVB, in more detail. The eVB application is configured to place a compiled version of the Survey application (DigitalSurvey in Figure 9.20) in \\Windows\Programs on your Windows CE-powered device.

The entire application will be based on three forms and two modules. The first form (frmDigitalSurvey listed in Listings 9.2 through 9.7) is in charge of providing all of the main menu options, the functions that will create the initial database on the Windows CE-based device, and will offer your user the option to synchronize the main SQL database with the new data collected in the Windows SQL CE database. As you can see on line 4 in Listing 9.2, eVB will automatically add a line of code that will close your application once you click the OK sign in the form's upper-right corner.

A very important part of the code in this form is line 10 in Listing 9.2, which invokes the function (lines 9 through 30 in Listing 9.17) InitREPL that will set all of the required parameters that will be needed later in the application development stage.

Lines 12 through 30 (Listings 9.3 and 9.4) will build the bottom menu bar. All of the options you would like to see in the main menu will be listed in this part of the code. Consider this: Whatever the top menu bar for VB is will be the bottom menu bar for eVB. Moreover, in VB, the building of the menu is an easy and

```
1. 'Form frmDigitalSurvey

2. Option Explicit
3. Private Sub Form_OKClick()
4. App.End
5. End Sub

6. Private Sub Form_Load()
7. Label1.Visible = False
8. CampaignID = "0"
9. StartMenuBar
10. InitREPL
11. End Sub
```

Listing 9.2 Startup form for sample application.

```
Private Sub StartMenuBar()
Dim mnuFile As MenuBarLib.MenuBarMenu
Dim mnuSurvey As MenuBarLib.MenuBarMenu
Dim mnuAbout As MenuBarLib.MenuBarMenu

'add File item
Set mnuFile = MenuBarControl.Controls.AddMenu("File", "mnuFile")
mnuFile.Items.Add 1, "mnuFileExit", "Exit"
mnuFile.Items.Add 2, "mnuFileAddSubscription", "Add Subscription"
mnuFile.Items.Add 3, "mnuFileSynchronize", "Synchronize"

'add Survey item
Set mnuSurvey = MenuBarControl.Controls.AddMenu("Survey",
    "mnuSurvey")
mnuSurvey.Items.Add 1, "mnuNewSurvey", "New Person"

'add About item
Set mnuAbout = MenuBarControl.Controls.AddMenu("About", "mnuAbout")
mnuAbout.Items.Add 1, "mnuNoOfRecords", "Records in DB"
End Sub
```

Listing 9.3 Menu bar definitions for main form.

intuitive task since there is a specific tool to do so. In eVB, all of the menu and submenu options are coded manually.

Lines 31 through 60 (Listings 9.5, 9.6, and 9.7) define the actions that will be taken once the menu items have been selected. Let's evaluate each listing:

- First is the Exit menu option (lines 34 and 35 in Listing 9.5). By executing the App.End statement, eVB code will terminate execution.
- The next part of the code (lines 36 through 44 in Listings 9.5 and 9.6) is something that you will only have to execute the first time you run this application. Note the part of the statement on line 37 (Listing 9.5) where data source is defined as survey.sdf ("data source=survey.sdf"). This part of the statement will define the name and location of the file to be created in the file system on the Windows CE device. You can adjust this line to place the local sdf database

```
Private Sub Label1_Click()
Label1.Visible = False
End Sub
```

Listing 9.4 End building the menu bar.

```
Private Sub MenuBarControl_MenuClick(ByVal Item As MenuBarLib.Item)
mnuItem = Item.Key
Select Case Item.Key

Case "mnuFileExit"
App.End

Case "mnuFileAddSubscription"
CEMerge.SubscriberConnectionString = "data source=survey.sdf"
```

Listing 9.5 Definition of menu actions.

in a different location (e.g., in the \\windows\programs directory where the compiled version of the application itself will reside).

- Code on lines 45 and 46 (Listing 9.6) will execute the code on lines 43 through 63 in Listing 9.18.
- Code on lines 47 and 48 (Listing 9.6) is nothing more than a pointer to load the form that will be used to collect a new survey.
- The last segment of the code in this form (lines 49 through 59 in Listings 9.6 and 9.7) is used to select count of all surveys taken and stored in the survey.sdf – tblSurveyResults table.

The SubscriberConnectionString property has many parameters that can be valuable during your future projects. The connection properties are listed in Table 9.7. It is clear that some of the options, from Table 9.7, will be used in the cases where we are storing financial, personal information, and other types of sensitive documents.

Note that on line 51 (Listing 9.6), we have explicitly provided the version of ADOCE control. If you do not specify the version, an earlier version of the control will be used. If there is no earlier version installed on your device, an error is returned.

NOTE

Window SQL CE databases can only be accessed with ADOCE 3.1 and later.

One of the options that you could add to the Survey application would be to compact the database. Invoking the CompactDatabase method can do this. The general syntax of this method would be:

CEMerge.CompactDatabase (SourceConnection, TargetConnection)

Table 9.7 Subscriber Connection String Parameters

PROPERTY NAME	MANDATORY/ OPTIONAL	SHORT DESCRIPTION
Provider	Optional	Defines the name of the data source provider.
Data Source	Mandatory	Defines the name of the database. Note that the standard file extension for this file type is .SDF.
Locale Identifier	Optional	Defines the collating order for string comparisons in the database. The default database locale is Latin1_General (0x00000409).
SSCE:Database Password	Optional	Defines the database password. In order to create a password-protected database, use this property and call method AddSubscription.
SSCE:Encrypt Database	Optional	If this option is set to TRUE, the newly created subscription database will be encrypted. Similar to the "Database Password" property in order to create an encrypted database, add this property to the connection string and call AddSubscription method from your code. Note that a database password must be specified in order to encrypt the database.
Sample string (in our chapter, the ADOCE object is called CEMerge): *CEMerge.SubscriberConnectionString = "Data Source=\survey.sdf; Locale Identifier=0x00000409;SSCE:Database Password={password};SSCE:Encrypt Database=TRUE"*		

```
On Error Resume Next
CEMerge.AddSubscription CREATE_DATABASE
If CEMerge.ErrorRecords.Count > 0 Then
ShowErrors CEMerge.ErrorRecords, "Add Subscription Failed"
Else
MsgBox "Subscription Added", vbOKOnly, "Add Subscription"
End If

Case "mnuFileSynchronize"
SynchronizeSurvey

Case "mnuNewSurvey"
frmNewSurvey.Show

Case "mnuNoOfRecords"
Dim oRS As adoce.Recordset
Set goADOcn = CreateObject("ADOCE.connection.3.1")
```

Listing 9.6 Definition of menu actions.

```
52. goADOcn.ConnectionString = gcstrLocalConnect
53. goADOcn.Open
54. Set oRS = goADOcn.Execute("select count( •) as TotalNumber from
        tblSurveyResults")
55. Label1.Visible = True
56. Label1.Caption = oRS.Fields("TotalNumber") & " Record(s)."
57. Label1.Refresh
58. oRS.Close
59. goADOcn.Close
60. End Select

61. End Sub
```

Listing 9.7 Number of records in tblSurveyResults.

This method can be used in two types of scenarios:

- To reclaim wasted space. This is important because disk space, or memory space in the case of Windows CE-powered devices, is still somewhat limited
- To repair corrupted databases.

There are a few details to keep in mind if you would like to use this method:

- If you consider the format of CompactDatabase method, you will notice that you cannot perform database compacting on the same database. You must specify the source connection string (connection string to your production database) and the target connection string (connection string to the compacted database). This implies that you need a database twice the size of your original for the compacting process to work.
- The source database must be closed before we attempt compacting.
- The target database must not exist before we attempt compacting.

Screens produced with the code for frmDigitalSurvey are shown in Figures 9.15 and 9.16. As you can see in Figure 9.16 (when you have executed part of the code that creates the initial subscription database), the number of records in the database will be 0. This screen will be shown to you when you have selected the Records in DB option from the main menu, shown in Figure 9.15. As we noted earlier, lines 49 through 59 in Listings 9.6 and 9.7 will be executed and the simple "Select count(*) from ...." statement will let us know how many records we collect that day. This information could be used, for example, if you are paid per survey, to calculate the amount that you have earned. The result set returned will be changed once we start entering data in our local copy of Windows SQL CE.

Figure 9.15 Application menu bar–records in database.

Because we have already created a local copy of the database, the next step would be to start collecting surveys. This task is accomplished through frmNewSurvey Form. Code for this part of the Survey application is provided in Listings 9.8 through 9.14.

The first event that will be executed is Form_Load (lines 31 through 37 in Listing 9.14). During the execution of this portion of the code, we will check to see if this is the first survey we are taking today by simply analyzing the content of global variable CampaignID on line 32 (Listing 9.14). If the value has not been assigned (the value in our application must be > 0), we will first present the end user with

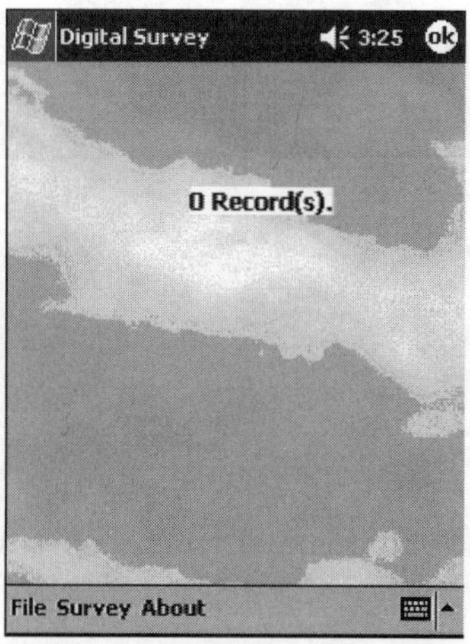

Figure 9.16 No records inside the tblSurveyResults.

Figure 9.17 Force campaign selection.

a notification message and then load the form that will allow the end user to select the type of survey to be performed. If the value of this global variable is equal to 0, then a message will be presented to the end user and the screen shown in Figure 9.17 will be displayed.

We will talk more about selecting the type of survey later in this chapter. For now, let's go back to frmNewSurvey.

The results of the survey are stored in tblSurveyResults as separate fields in the database (fields Question1, Question2, and so on). For example, the default value for the answer to the first question in the survey (Age question)

```
' Form frmNewSurvey
Option Explicit

Private Sub Command1_Click()
Dim Q1, Q2, Q3, Q4, Q5
Q1 = CStr(Abs(CInt(Option1.Value))) &
   CStr(Abs(CInt(Option2.Value))) & CStr(Abs(CInt(Option3.Value)))
Q2 = CStr(Abs(CInt(Option4.Value))) &
   CStr(Abs(CInt(Option5.Value)))
Q3 = CStr(Abs(CInt(Option6.Value))) &
   CStr(Abs(CInt(Option7.Value))) & CStr(Abs(CInt(Option8.Value)))
Q4 = CStr(Abs(CInt(Option9.Value))) &
   CStr(Abs(CInt(Option10.Value))) & CStr(Abs(CInt(Option11.Value)))
Q5 = CStr(Abs(CInt(Option12.Value))) &
   CStr(Abs(CInt(Option13.Value))) & CStr(Abs(CInt(Option14.Value)))
```

Listing 9.8 Code for starting a new survey.

```
10. If (Replace(Q1, "0", "") + Replace(Q2, "0", "") + Replace(Q3, "0",
      "") + Replace(Q4, "0", "") + Replace(Q5, "0", "")) <> "11111" Then
11. MsgBox "You have missed one field", vbOKOnly, "Missed Field"
12. Exit Sub
13. End If
```

Listing 9.9 Code to check if field is missed.

is "000", since default values for radio buttons are 0. Now, if we select the 15–30 radio button, the resulting string will be "100", and if we select 31–45, the resulting string will be "010". You can find the code that will calculate the value for question one on line 5 in Listing 9.8.

One of the problems that we have to resolve is to make sure that all of the questions are answered. To accomplish this, we will determine that if the question is answered, it must have at least one number 1 and a string of zeros. If we replace all of the zeros in all of the answers and add the strings together, we should end up (if all of the questions are answered) with string 11111. If one question is missed, then the string will be "1111". Lines 10 through 13 in Listing 9.9 will provide us with a method of detecting if any of the questions are missed. If some of the questions are not answered when we click Save, the screen shown in Figure 9.18 will appear.

We will not know exactly which question is not answered (we could with some minor code adjustments), but the end user will not be allowed to save the incomplete survey.

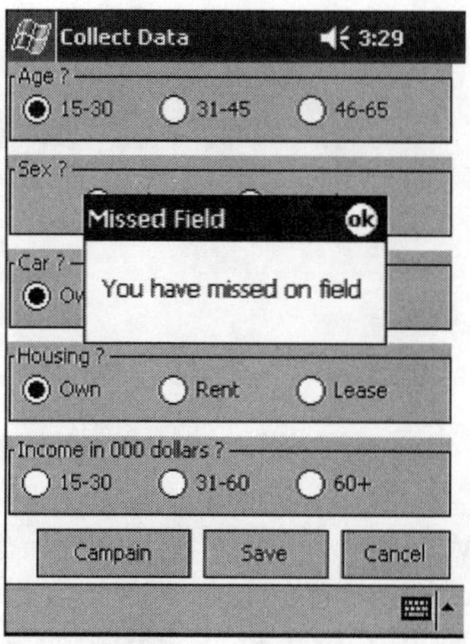

Figure 9.18 Survey form.

If the person completing the survey made sure that all the questions were answered, then lines 14 through 22 in Listing 9.10 will be executed. We will first open connection to our survey.sdf database (lines 15 and 16 on Listing 9.10), and then execute the SQL Insert statement that will create a record in our tblSurveyResults table.

```
14. Dim oRS As adoce.Recordset
15. Set goADOcn = CreateObject("ADOCE.connection.3.1")
16. goADOcn.ConnectionString = gcstrLocalConnect
17. goADOcn.Open
18. goADOcn.Execute ("insert tblSurveyResults (SurveyID, Question1,
      Question2, Question3, Question4, Question5, Upload) values ('"
      & CampaignID & "','" & Q1 & "','" & Q2 & "','" & Q3 & "','" &
      Q4 & "','" & Q5 & "','N')")
19. goADOcn.Close
20. CleanUp
```

Listing 9.10 Insert new record in tblSurveyResults.

Line 20 in Listing 9.10 will run the CleanUP function that is part of the Other-Global module. Cleanup in this case means that all of the radio buttons are reset to the unselected state and prepared for a new entry. Code in Listing 9.11 will close the subroutine started in Listing 9.8.

Note that we are inserting a value of "N" for the Upload field (line 18 in Listing 9.10). Take a look back for the moment at the WHERE clause in the SELECT statement we created for deciding what data to filter:

```
21. frmNewSurvey.Refresh
22. End Sub
```

Listing 9.11 Prepare for new record.

```
SELECT<published_columns> FROM [dbo].[tblSurveyResults] WHERE
   tblSurveyResults.Upload = 'Y'
```

Yes, rows of data will be replicated to the Windows SQL CE database, but only if the value of the "Upload" field is "Y". Otherwise, data will be propagated one way, to the central Windows SQL database, which is exactly what we want.

```
23. Private Sub Command2_Click()
24. Me.Hide
25. frmDigitalSurvey.Show
26. End Sub
```

Listing 9.12 Show survey form.

Lines 23 through 26 in Listing 9.12 will simply hide the current form and take us back to the main menu form.

```
27. Private Sub Command3_Click()
28. Me.Hide
29. frmCampaign.Show
30. End Sub
```

Listing 9.13 Show campaign selection form.

In case we would like to move between different types of surveys throughout the day, we can invoke a campaign selection form (frmCampaign) by executing lines 27 through 30 in Listing 9.13. It is also important to ensure that our users select campaign when they start the application. Code in Listing 9.14 will check to see if any campaign has been selected to that point (line 32), and if it is not, it will raise a message box on the screen (line 33) that tells the users to select the campaign.

```
31. Private Sub Form_Load()
32. If CampaignID = "0" Then
33. MsgBox "Please select valid Campaign", vbOKOnly, "Select Campaign"
34. Me.Hide
35. frmCampaign.Show
36. End If
37. End Sub
```

Listing 9.14 Code to select current campaign.

The Campaign Selection form is driven from the tblSurveys table, and unlike the tblSurveyResults, we do not perform any filtering of data. All of the rows created on the main Windows SQL server database will be published to the Windows SQL CE subscription database. The code behind the form shown in Figure 9.19 does two simple things:

- Populates the drop-down box for the selection of campaign type
- Enables us to return to the data collection screen

```
1. 'Form frmCampaign

2. Option Explicit

3. Private Sub Command1_Click()
4. CampaignID = Right(Combo1.Text, 1)
5. Me.Hide
6. frmNewSurvey.Show
7. End Sub
```

Listing 9.15 Return to Survey Collection screen.

The code on lines 3 through 7 in Listing 9.15 will resolve the need to return to the data collection screen by hiding itself (line 5 in Listing 9.15) and showing the survey data collection form (frmNewSurvey on line 6 in Listing 9.15).

Code on lines 8 through 22 in Listing 9.16 will query the table tblSurveys to present the end user with a list of available surveys by simply selecting all of

```
8. Private Sub Form_Load()
9. Dim oRS As adoce.Recordset
10. Set goADOcn = CreateObject("ADOCE.connection.3.1")
11. goADOcn.ConnectionString = gcstrLocalConnect
12. goADOcn.Open
13. Set oRS = goADOcn.Execute("select        • from tblSurveys order
      by SurveyID")
14. Do While Not oRS.EOF
15. Combo1.AddItem oRS.Fields("SurveyTitle") & "," & oRS.Fields
      ("SurveyID")
16. oRS.MoveNext
17. Loop
18. Combo1.Text = "Select Here"
19. Combo1.Refresh
20. oRS.Close
21. goADOcn.Close
22. End Sub
```

Listing 9.16 Select and show available surveys.

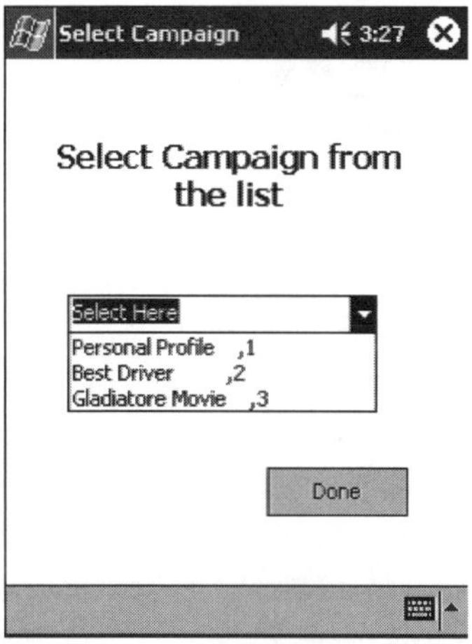

Figure 9.19 Campaign Selection screen.

the available rows in that table. When the end user selects the desired survey from the drop-down box and clicks the Done button, the value of the global variable CampaignID will be assigned or changed depending on the stage at which we performed this action.

At this point, four more steps are required before we complete our sample application and confirm that it works as planned. First, we have to start our application on the Windows SQL CE-powered device and install all of the required components. We wrote in more detail about the subject of Integrated Development Environment (IDE) in Chapter 7. We will note that by selecting the Runtime Files and "Project Components" checkboxes on the Project Properties screen (see Figure 9.20) of your eVB and subsequently executing the code by pressing the F5 function key, you will effectively compile the code, deliver the compiled version of the application to the required directory (in this case, \Windows\Programs), and deliver all of the required runtime files and components.

When the copy and installation is complete, the Survey application should start immediately on your Windows CE-powered device. At this point we can create an initial subscription database by selecting File->Add Subscription from the main application menu.

As soon as the subscription has been added, we can add new records to the database (do not forget to select the type of the campaign you will be running, which is in this case the "Personal Profile" survey). Let's now start our application and add couple of records by completing the survey. These two records will emulate a busy surveyor's day on the streets.

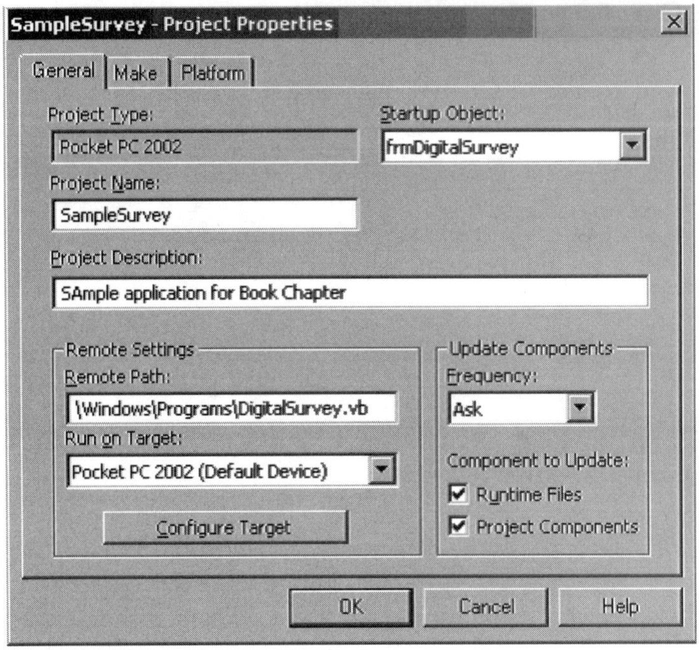

Figure 9.20 eVB Survey project properties.

We are now back at the office, and the final task of the day (as well as the goal of this sample) is to upload collected data to the central database, download currently active Surveys, and clean up the database on the Windows CE-powered device. Before you synchronize the data, let's see how many records we have stored in the local SQL CE database. Execute About->Records in the database (DB) option from the main application form and you should see that we have two records.

Next, we will execute File->Synchronize from the main application form menu. After a couple of seconds of network and disk activity, the user should see the screen shown in Figure 9.21. The message on the screen tells us that two records have been added to the tblSurveyResults on the main Windows SQL server, and that two records have been deleted from the same table on the Windows SQL CE database.

In order to confirm this, we will again select About->Records in DB and we should see a message on the screen saying that no records were found. There should be no records at this point in our Windows SQL CE database, and we are now ready for the next day.

Listings 9.17 and 9.18 are eVB modules that deal with common functions used throughout the sample application.

Figure 9.21 Synchronization results.

```
' Module OtherGlobal
Option Explicit

Dim mnuItem
Dim CEMerge As SSCE.Replication
Public goADOcn As adoce.Connection
Public goADOrs As adoce.Recordset
Public CampaignID
Const gcstrLocalConnect = "Provider=Microsoft.SQLServer.OLEDB
    .CE.1.0;Data Source=Survey.SDF"

Sub InitREPL()
 Set CEMerge = CreateObject("SSCE.Replication.1.0")
 CEMerge.InternetURL = "http://{your server
    name}/anon/sscesa10.dll"
 CEMerge.InternetLogin = ""
 CEMerge.InternetPassword = ""
 CEMerge.Publisher = "{your server name}"
 CEMerge.PublisherDatabase = "survey"
 CEMerge.Publication = "chapter"
 CEMerge.PublisherSecurityMode = DB_AUTHENTICATION
 CEMerge.PublisherLogin = "sa"
 CEMerge.PublisherPassword = ""
 CEMerge.PublisherNetwork = DEFAULT_NETWORK
 CEMerge.Subscriber = "Srdjan"
 CEMerge.SubscriberConnectionString = "data source=survey.sdf"
 CEMerge.Distributor = "{your server name}"
 CEMerge.DistributorNetwork = DEFAULT_NETWORK
 CEMerge.DistributorSecurityMode = DB_AUTHENTICATION
 CEMerge.DistributorLogin = "sa"
```

Continues

Listing 9.17 eVB module with synchronization code.

```
CEMerge.DistributorPassword = ""
CEMerge.ExchangeType = BIDIRECTIONAL
CEMerge.Validate = NO_VALIDATION
End Sub

Sub ShowErrors(ErrColl As SSCEErrors, strCaption As String)
Dim ErrRec As Object    'SSCE.ErrorRecords
Dim strErr As String
strErr = ""
For Each ErrRec In ErrColl
strErr = strErr & "Source: " & ErrRec.Source & vbCrLf
strErr = strErr & "Number: " & Hex(ErrRec.Number) & vbCrLf
strErr = strErr & "NativeError: " & ErrRec.NativeError & vbCrLf
strErr = strErr & "Description: " & ErrRec.Description & vbCrLf &
    vbCrLf
Next ErrRec
MsgBox strErr, vbOKOnly, strCaption
End Sub
```

Listing 9.17 eVB module with synchronization code. *(Continued)*

Now, we have to explain the piece of code that handles synchronization. This is achieved with the code on lines 43 through 63 in Listing 9.18, but it could be simplified to only three lines of code. Lines 46, 51, and 61 will do the majority of the work, and the rest of the code is just simple error handling. In short, the sequence of synchronizing the database is Initialize->Run->Terminate.

```
43. Sub SynchronizeSurvey()
44. Dim str As String
45. On Error Resume Next
46. CEMerge.Initialize
47. If CEMerge.ErrorRecords.Count > 0 Then
48. ShowErrors CEMerge.ErrorRecords, "Initialization Failed"
49. Else
50. On Error Resume Next
51. CEMerge.Run
52. If CEMerge.ErrorRecords.Count > 0 Then
53. ShowErrors CEMerge.ErrorRecords, "Synchronization Failed"
54. Else
55. str = "Synchronization Complete" & vbCrLf
56. str = str & "Publisher Changes = " & CEMerge.PublisherChanges &
       vbCrLf
57. str = str & "Publisher Conflicts = " & CEMerge.PublisherConflicts
       & vbCrLf
58. str = str & "Subscriber Changes = " & CEMerge.SubscriberChanges &
       vbCrLf
```

Listing 9.18 Execute synchronization and cleanup code.

```
MsgBox str, vbOKOnly, " S Y N C H R O N I Z E "
End If
CEMerge.Terminate
End If
End Sub

Sub CleanUp()
frmNewSurvey.Option1.Value = False
frmNewSurvey.Option2.Value = False
frmNewSurvey.Option3.Value = False
frmNewSurvey.Option4.Value = False
frmNewSurvey.Option5.Value = False
frmNewSurvey.Option6.Value = False
frmNewSurvey.Option7.Value = False
frmNewSurvey.Option8.Value = False
frmNewSurvey.Option9.Value = False
frmNewSurvey.Option10.Value = False
frmNewSurvey.Option11.Value = False
frmNewSurvey.Option12.Value = False
frmNewSurvey.Option13.Value = False
frmNewSurvey.Option14.Value = False
End Sub
```

Listing 9.18 *(Continued)*

The last module (Listing 9.19) is a collection of common Windows SQL 2000 CE-related constants.

```
'Module ssceconsts - common Windows SQL CE constants

Const DB_AUTHENTICATION = 0
Const NT_AUTHENTICATION = 1
Const DEFAULT_NETWORK = 0
Const TCPIP_SOCKETS = 1
Const MULTI_PROTOCOL = 2
Const CREATE_DATABASE = 1
Const EXISTING_DATABASE = 0
Const ATTACH_SUBSCRIPTION = 3
Const LEAVE_DATABASE = 0
Const DROP_DATABASE = 1
Const NO_VALIDATION = 0
Const ROWCOUNT_ONLY = 1
Const BIDIRECTIONAL = 3
Const TRACKINGON = 1
Const TRACKINGOFF = 0
Const ENCRYPTCOPY = 0
Const ENCRYPTOFF = 1
Const ENCRYPTON = 2
```

Listing 9.19 Windows SQL 2000 CE required constants.

The sample application presented in this chapter is only an example and was not intended to provide a production-level application. To achieve the quality required for the production level, some changes and additions will have to be performed. For example:

- You would, most likely, want to create a completely database-driven frmNewSurvey form that will dynamically "paint" the screen as soon as a new survey type is selected.
- A table and the code that will hold statistical data with numbers about surveyors' efficiency will have to be created
- Much more robust error handling will have to be put in place.

These are only some of the ideas that will improve our sample application.

Final Thoughts

Microsoft did an excellent job with the Windows SQL Server 2000 CE edition. What separates this product from similar products is that even less experienced DBA and application developers can start application development with a minimum amount of effort spent on system administration, but that is the strength of all Microsoft products.

On the other hand, Microsoft SQL Server is "bound" to run on a Windows-based platform. In other words, if your company (or your project) is completely based on Microsoft's line of products, then this is the product to use. However, if you have to cover a variety of portable devices (Palm, Windows CE, and others) and a variety of legacy systems with your application, you will have additional choices that we will talk more about in the next two chapters.

CHAPTER 10

Sybase

This chapter deals with one of the leading database management systems in the world. Not only is Sybase a leading database on the Unix and NT platforms, they are also a leader in the portable wireless arena.

Before we begin, we'd like to recognize and sincerely thank Sybase as a major contributor to and sponsor of our book. Of all the major database vendors we contacted, Sybase's customer service and marketing department responded quickly and with professional acknowledgment of our project and of the wireless industry as a whole. The dedication of Sybase to the wireless industry, enterprise integration, and extending corporate investments and assets to the mobile portable world is formed from a true spirit of partnership and client interest. Hats off to Sybase and the staff at iAnywhere.

Sybase provides excellent manuals for all the features presented in this chapter, and then some.

Overview

Sybase is a big player in the database world with headquarters in Oakland, California, and their wireless solution center, iAnywhere solutions, located in Waterloo, Ontario, Canada. This division was formerly a firm called Watcom, which was created by the minds at Waterloo University. Watcom was basically a competitor to Microsoft's Access database software before being acquired by

Sybase. Their focus was then shifted to the wireless industry, as we understand it. They became one of the first major DBMS vendors to focus efforts on small portable wireless devices.

This chapter gives a brief glimpse into Sybase's Adaptive Server Anywhere (ASA) database on a Windows 2000 server with connectivity to the popular Adaptive Server Anywhere UltraLite database. Adaptive Server Anywhere is an extension to Sybase's Adaptive Server Enterprise (ASE), which is primarily used in the corporate enterprise. As a quick insight, many applications written on ASE can migrate to ASA.

Sybase has ASE for the corporate world primarily for Unix and NT. ASA is a smaller version, which in itself is becoming quite popular. This version is competitive to Microsoft's Access database. ASA has a stronger, more advanced version of SQL (more features) and is mostly found in smaller, independent applications. UltraLite goes a step further in the smaller category. UltraLite is a smaller version of the ASA technology and is used in building database applications for small devices, such as handheld organizers and PDAs. To connect and synchronize these two databases on the two distinct platforms, NT and the mobile, Sybase's iAnywhere division created MobiLink. This middleware component is a flexible area that allows administrators to pick and choose data and gives them the means to synchronize specific data between databases. More information on these will be provided throughout this chapter. For now, Figure 10.1 shows the overall Sybase approach.

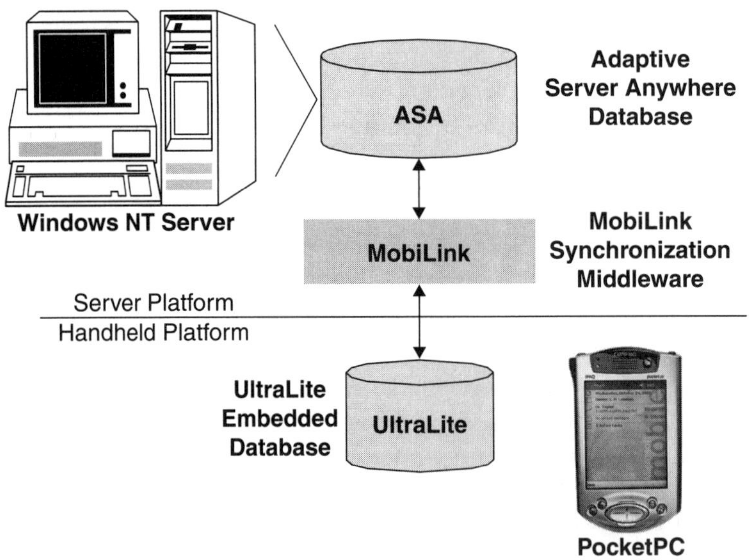

Figure 10.1 iAnywhere Mobile overview.

From the application point of view, several components (shown in the following list) are required. The Sybase components are bundled together in Sybase SQL Anywhere Studio 8.

- Microsoft Windows NT Server (we'll be using Windows 2000)
- Sybase SQL Anywhere Studio 8
  - Sybase Adaptive Server Anywhere
  - Sybase MobiLink
  - Sybase UltraLite
- MobileBuilder 2.1—application builder
- Microsoft ActiveSync 3.5 (minimum version is 3.1—delivered with Compaq iPaq Pocket PC)
- Pocket PC—Compaq iPAQ H3870

The following would be minimum PC requirements:

- 128MB of RAM (256MB recommended)
- Pentium III CPU "ticking" at 500 MHz minimum
- 200–300MB of free hard disk space
- Available USB or RS-232 (usually known as Com port)

SQL Anywhere Studio

This chapter focuses on Sybase's SQL Anywhere Studio version 8. It is Sybase's mobile strategy, or should we say Sybase's enterprise extension strategy, since it allows the current enterprise data and applications to cross over to the wireless mobile world.

The concept is simple: Allow the current systems, databases, and applications to function as usual, and the SQL Anywhere Studio components will allow the extension of the information and functionality to the mobile workforce. To do this, there are basic areas and issues to overcome, such as how to send and receive data from the mobile devices and how to synchronize it into and from the database on the enterprise.

There are three basic main components in Sybase's strategy: A relational database on the enterprise, another on the mobile device, and a data replication and synchronization methodology. Of course, there must be some type of database manipulation component in a simple, easy-to-use fashion. This is accomplished with the Sybase Central administration tool.

There's much more to Sybase's overall strategy than this chapter explains or uses. We'll only scratch the surface to show you what the Sybase components are and how they fit together. Now let's look at each component.

Relational Database Components

Adaptive Server Anywhere is the transaction-based SQL relational database at the heart of SQL Anywhere Studio. It's designed for personal and enterprise use and is the core of many delivered and developed applications. Adaptive Server Anywhere runs on a wide range of operating systems, including many flavors of Windows and Unix, and on Novell NetWare, as well. It runs on hardware ranging from multiple-CPU workgroup servers to the most modest PCs, as well as on Windows CE devices.

UltraLite is the solution for relational database applications on small devices, such as the Palm Computing Platform, Windows CE, and VxWorks. It has a very small footprint with less than 50KB of disk space on devices, such as PocketPC. UltraLite has synchronization capabilities from its built-in MobiLink component.

Data Synchronization and Replication

Sybase has several data synchronization and replication strategies depending on requirements and usage. They are MobiLink, SQL Remote, and Replication Agent. The following explains each of these in Sybase's own words directly from their manual.

MobiLink. For two-way synchronization of data between a central database and many remote UltraLite or Adaptive Server Anywhere databases. The central database can be Adaptive Server Anywhere, Adaptive Server Enterprise, or another DBMS such as Oracle, Microsoft SQL Server, or IBM DB2.

ActiveSync. Synchronization software for Microsoft Windows Powered Pocket PCs. Adaptive Server Anywhere MobiLink clients can use ActiveSync version 3.1 or 3.5. ActiveSync governs synchronization between a Windows CE device and a desktop computer. A MobiLink provider for ActiveSync governs synchronization to the MobiLink synchronization server.

SQL Remote. For two-way, message-based synchronization replication of data between a central database and many remote databases. With SQL Remote, you can replicate data between laptop computers and a central database, using email or dial-up access.

Replication Agent. For replicating data from Adaptive Server Anywhere databases to other databases via Sybase Replication Server.

Database Administration Tools

Every database needs some type of administration tool. Of course, one could always perform all administration via command lines, but that's long, boring, and very difficult because you must remember each command with absolute error-free syntax. The last thing most database administrators would want to do is maintain a database via individual commands. Therefore, a nice GUI database management tool is a definite requirement for setup and ongoing maintenance.

Sybase Central. SQL Anywhere Studio includes the Sybase Central database management tool, shown in Figure 10.2. Notice that the Utilities folder contains many types of database management utilities. The place to start would obviously be with the *Create Database* option. Click on >Start >Programs >Sybase SQL Anywhere 8 >Sybase Central. This will invoke the application you'll see throughout this chapter. More on how to build a database later in the chapter.

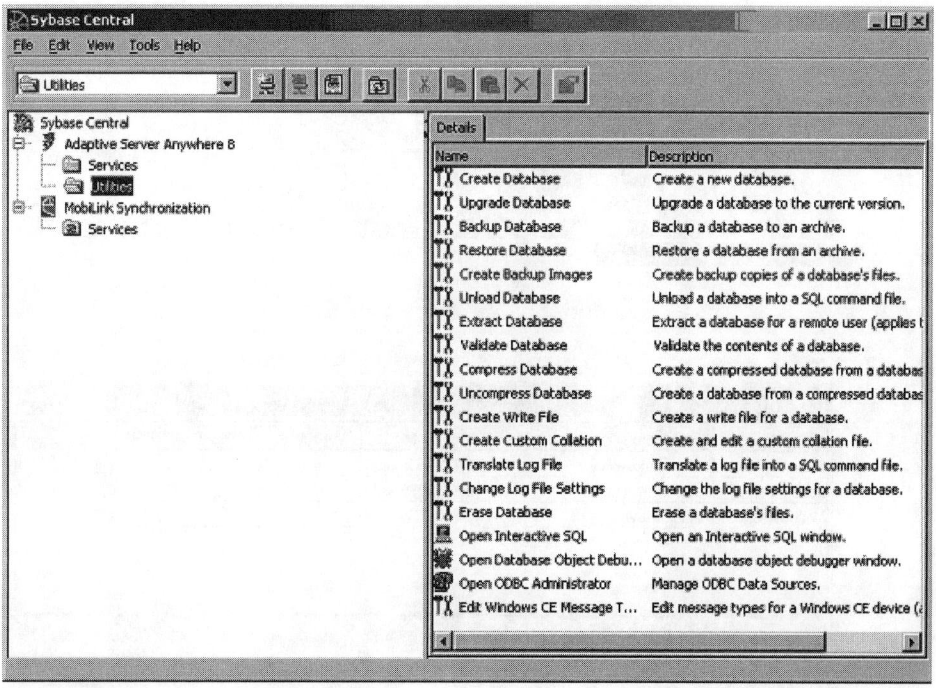

Figure 10.2 Sybase Central.

Interactive SQL. This is an added tool from the Utilities folder, shown in Figure 10.2. It's used to query and manipulate data in the database. It has been the backbone to Sybase's database since Sybase and Microsoft SQL Server were one and the same. ASA is a scaled-down version of ASE, which was originally called SQL Server when Microsoft and Sybase were partners in Windows-based database development. Microsoft kept the SQL Server name and Sybase moved toward ASE and both developed their engines and relational database management systems in their own vision. Figure 10.3 is a sample of what the tool looks like and a typical select statement.

Embedded Database Architecture

You can build an application and database on a single computer. In its simplest form, this is a standalone application as it is self-contained, with no connection to other databases. In this case, it is common to refer to the database as an *embedded database* because, as far as the end user is concerned, the database is a part of the application. When a database server is used as an embedded database, it is sometimes referred to as a *database engine*.

Many relational database management systems require experienced database administrators (or sometimes programmers) for administration. A major pos-

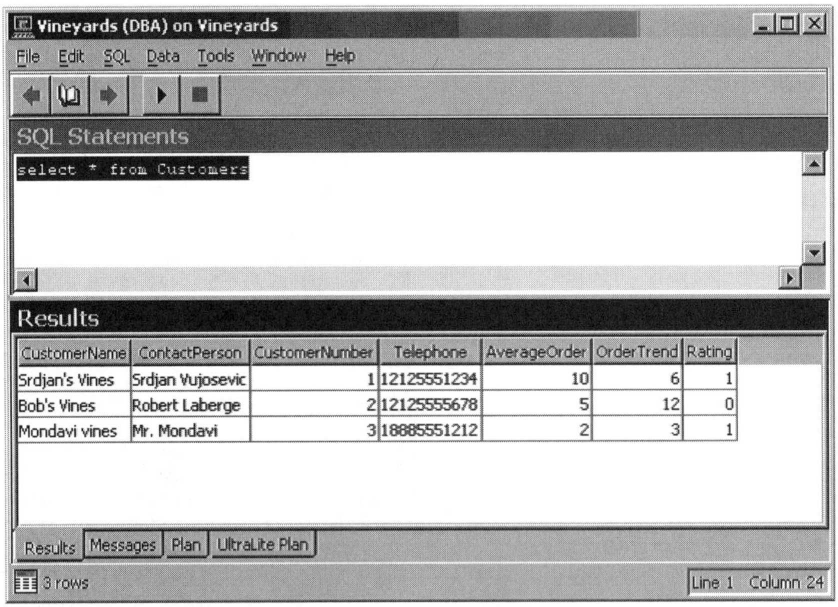

Figure 10.3 Interactive SQL.

itive point about embedded databases is the capability to run entirely without administration—standalone.

The ASE runs as a separate database management system with applications connecting and inquiring directly to the database from a distinct executable. However, the ASA personal database server is generally used for standalone applications. A client application connects through a programming interface to a database server running on the same machine.

For the UltraLite deployment strategy for small devices, as in our case, the database and application are tightly coupled, forming an embedded database or standalone application. The developer is very specific about which tables are required, the number of rows (data), and wraps the database and application around each other forming a single system or process, which can then be easily deployed to the mobile device.

This is a key point in small device application development. Since the devices are small in size, disk space (generally), and processing power, a major objective of building mobile applications is to keep the footprint of the database as small as possible. Embedded databases allow for this strategy to be deployed.

Client-Server Architecture

Referring back to Chapter 3, "Client-Server Architecture," it is important to understand that extending the enterprise or desktop applications and databases to small mobile devices is a matter of configuring a proper client-server architecture.

Since the small devices are, for lack of a better explanation, small, the developer must ensure a proper architecture to keep the units as thin as possible. This means reducing any overhead on the devices. Only deploy data and application components absolutely essential to the operations of the application. This becomes even more critical when creating extensions to data warehouses on the small devices. In other words, prepare everything in advance on the enterprise or desktop, and deploy as little as possible.

Synchronization becomes an issue if too much data is being transferred or if the application is too large. Downloading (or uploading) takes time, as we're all aware when we try to download a file from the Internet via our home telephone lines. If the mobile unit is connecting to the desktop via a wireless modem, the same issues apply. If connecting via a USB port, the transfer rate is much higher, but the architecture issues still apply.

Of course, the application on the client is designed to read and manipulate server and/or new data and its efficiencies are within the developers' hands, but data is the key. The information must be properly configured on the server before deployment to the client or clients. How will the server data be replicated, partitioned, and organized? Will the data be cached or will it constantly be queried from end to end? Many issues exist, some small and some all-encompassing. The point is to be aware of the architecture including a deployment strategy.

You can use Adaptive Server Anywhere to build an installation with many applications, running on different machines, connected over a network to a single database server running on a separate machine. This is a *client-server* environment, where the interface library is located on each client machine. In this case, the database server is the Adaptive Server Anywhere network database server, which supports network communications and is called a *multi-user database*. No changes need to be made to a client application for it to work in a client-server environment, except to identify the server to which it should connect. However, when extending ASA to UltraLite for small devices, the client environment does change. A mapping strategy on the server before deployment must be in place where the data is mapped from ASA to the UltraLite version.

Adaptive Server Anywhere

Let's look at the first piece of the Sybase puzzle, the database on the enterprise, or, in our case, on the desktop called ASA.

Overview

For many years, Sybase Adaptive Server Anywhere has provided relational database technology designed specifically for the needs of workgroup (enterprise), mobile, and embedded computing. The product has been designed from the ground up with this market in mind (again we'll quote directly from the ASA manuals, as Sybase says this better than we do):

- Adaptive Server Anywhere is designed to operate efficiently with limited memory, CPU power, and disk space. Core features such as the query optimizer and the data caching mechanism are designed specifically to operate without extravagant use of resources. At the same time, Adaptive Server Anywhere contains the features needed to take advantage of workgroup servers, including support for many users, scalability over multiple CPUs, and advanced concurrency features.
- Adaptive Server Anywhere is a cross-platform solution. The same database runs on Windows (Windows 95 and its successors, Windows NT and its successors, and Windows CE), Unix, including Linux, and Novell NetWare. You can move a database file from one operating system to another.

- Adaptive Server Anywhere is designed to operate without administration, making it ideal for use as an embedded database. Adaptive Server Anywhere provides a self-tuning query optimizer, built-in scheduling and event-handling capabilities, as well as autostart and autostop mechanisms.
- Many years of experience working with successful customers have led to a rich set of field-tested features. Not only the standard checklist features of stored procedures, triggers, declarative referential integrity, full transaction processing, and recovery, but all the little extras that can make the difference between a successful project and a failure.
- SQL Anywhere synchronization technologies (SQL Remote and MobiLink) mean that you can integrate Adaptive Server Anywhere databases into your organization's infrastructure.

DBMS Specifics

Adaptive Server Anywhere is a relational database system for use in applications on workgroup servers, desktop and laptop computers, on Windows CE devices, and in embedded systems. The following are many interesting points about ASA:

- The database can be up to several gigabytes in size.
- Can manage single-user databases in as little as 3MB of memory.
- Has reached an installed base of over 6 million customers.
- ASA version 8 follows SQL/99 standards for clarifying the use of aggregate functions when they appear in subqueries. ASA complies completely with the SQL-92-based United States Federal Information Processing Standard Publication (FIPS PUB) 127. ASA is entry-level compliant with the ISO/ANSI SQL-92 standard, and with minor exceptions, is compliant with SQL-99 core specifications.
- ASA provides a native ODBC 3.5 driver for high performance from ODBC applications, and an OLE DB driver for use from ActiveX Data Object (ADO) programming environments. It comes with Sybase jConnect for JDBC, and supports Embedded SQL and Sybase Open Client interfaces.
- Can be run on many operating systems, including Windows 95/98 and ME, Windows NT, Windows 2000 and Windows XP, Windows CE, Novell NetWare, Sun Solaris, and Linux.
- Supports TCP/IP network protocol and the SPX protocol for Novell NetWare.

ASA does not support all ASE components. For specific details, you'll have to refer to the manuals, as there are many discrepancies. For example, for those familiar with ASE, a master database exists to hold all metadata about all databases on the server. However, for ASA, each database is an independent entity, containing all of its own system tables. When a user connects, he or she connects to an individual database, not a server. There is no system-wide set of system tables maintained at a master database level. Each ASA database server can dynamically load and unload multiple databases, and users can maintain independent connections on each. ASA manages its own resources automatically, and its databases are regular operating system files.

In ASE, the database owner (user ID dbo) owns the system tables. In ASA, the system owner (user ID SYS) owns the system tables. A dbo user ID also owns the ASE-compatible system views provided by ASA.

The database administrator (DBA authority) has, like the ASE database owner, full permissions on all objects inside the database (other than objects owned by SYS), and can grant other users the permission to create objects and execute commands within the database. The default database administrator is user ID DBA.

In ASA, user IDs and passwords follow the case sensitivity of the data. The default user ID and password for case-sensitive databases are uppercase DBA and SQL, respectively.

Operating Requirements

Operating system and CPU. You must have one of the following in order to run the Adaptive Server Anywhere database server:

Windows NT. Version 4.0 or later, or Windows 2000, or Windows XP. The documentation describes the use of Windows NT. SQL Anywhere Studio components that run on Windows NT also run on Windows 2000 or Windows XP.

Windows. Windows 95, Windows 98, or Windows ME.

Windows CE. Versions 2.11 and 3.0 are supported. The OLE DB driver on CE is available only for the MIPS, ARM, and SH3 chips. The operating system is supported on the following processors:

- MIPS processor
- Hitachi SH3 or SH4 processor. (both for version 2.11, but only SH3 for 3.0)
- ARM processor (our example Compaq iPAQ H3870 uses the ARM SA1110 CPU)

Novell NetWare. Version 3.2, 4.11, 4.2, or 5.x.

Unix, including Linux. For a detailed list of supported Unix operating systems, see the *SQL Anywhere Studio Read Me First for UNIX*.

PC hardware. For PC operating systems, an Intel 486 or higher CPU, or compatible CPU, is required.

Memory. Adaptive Server Anywhere can run with as little as 3MB of memory. If you use Java in the database, Adaptive Server Anywhere requires 8MB of memory. Your computer must have this much memory in addition to the requirements for the operating system.

UltraLite

The second piece of the Sybase small device deployment strategy is the UltraLite methodology.

Overview

Small computing devices such as handheld computers, pagers, and mobile phones create a demand for databases with even more modest memory requirements than Adaptive Server Anywhere. An obvious option would have been to produce a trimmed-down relational database engine, but Sybase found that each application had a distinct set of essential features unique to that specific application. Further, such an approach would mean that developers would have to learn two different databases, inevitably different in some ways.

Instead, the UltraLite strategy was developed (see Figure 10.4). It involves using a reference database and application source code to generate a relational database engine containing only those features of ASA used by the application. Each query is stored with a complete access plan for fast execution; the code needed to execute just the required application steps is built directly into the UltraLite database engine. You define the data access requests that an UltraLite application can carry out by adding a set of SQL statements to the UltraLite project for that application in your reference database. The UltraLite generator then creates the code for a database engine that can execute the set of SQL statements. Each UltraLite database engine is different, but many are only a few tens of kilobytes, and can easily be run on a small device.

ASA serves as a reference database when you build your UltraLite application, and so your SQL statements, data types, and so on, are exactly those of ASA. When you add SQL statements to a reference database, you assign them to an UltraLite *project*. UltraLite is a deployment technology for Adaptive Server Anywhere, not a different database. The tasks that each UltraLite database engine is built to perform are carried out in a manner completely compatible with Adaptive Server Anywhere.

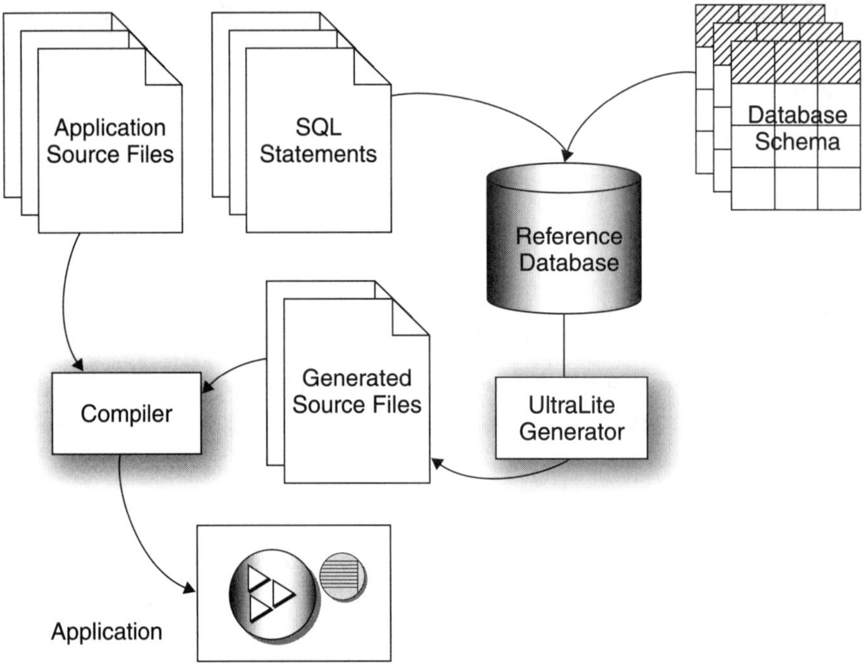

Figure 10.4 UltraLite process (from Sybase documentation).

UltraLite supports the Palm Computing Platform, Windows CE, and other operating systems used in small devices, such as Java and VxWorks.

UltraLite Architecture

SQL database products typically use a client-server architecture. The database is normally stored in one or more files in the file system, or directly on a storage device. The database server receives and processes SQL requests from client applications, maintaining the database according to the requests. The server protects you from failures by guaranteeing that complete transactions are recovered in case of a failure and incomplete transactions are rolled back.

UltraLite has the same client-server architecture as other SQL database systems. However, the UltraLite database engine is not a separate process, but is instead a library of functions that is called by an application. If you build your application using C/C++, this engine can be accessed either as a DLL or from a statically linked library. If you build your application in Java, the engine is accessed from Java byte code stored in a JAR file.

C/C++ Deployment

If you build your application using C/C++, the UltraLite development tools generate C/C++ source code that is compiled along with your application source code. When you link your application, you link all of the compiled C/C++ together with the UltraLite runtime library or imports library. The result is a single executable file containing application logic and database logic required by the application.

When it is first executed on a new device, this executable automatically creates the UltraLite database for your application. This database is initially empty, but you can add data, either explicitly (with insert commands) or through synchronization with a central database.

Java Deployment

If you build a Java UltraLite application, the UltraLite development process generates Java source code that represents the database schema. The generated source file is compiled into classes that you deploy as part of your application with the UltraLite runtime JAR file. You might want to package all files together into a single JAR file for ease of deployment.

When it is first executed on a new device, the UltraLite runtime automatically creates the database for your application. This database is initially empty, but you can add data, either explicitly or through synchronization with a central database.

Persistent Memory

UltraLite provides protection against system failures. Some UltraLite target devices have no disk drive, but instead feature memory that retains content when the device is not running. The storage mechanism for the UltraLite database is platform dependent, but is managed by the UltraLite runtime library, and does not need explicit treatment from the application developer.

Fixed Schema

UltraLite does not allow the schema of an UltraLite database to be modified when the application is deployed. When a newer version of the application requires more tables or more columns, the newer version of the application is deployed and the UltraLite database is repopulated through synchronization.

The SQL statements used in the application must be determined at compile time. In other words, you cannot dynamically construct a SQL statement within an UltraLite application and execute it. However, SQL statements in UltraLite applications can use placeholders or host variables to adjust their behavior.

UltraLite Features

UltraLite is a deployment technology for Adaptive Server Anywhere databases, aimed at small, mobile, and embedded devices. Intended platforms include cell phones, pagers, and personal organizers.

UltraLite provides the following benefits for users of small devices:

- The functionality and reliability of a transaction-processing SQL database
- The ability to synchronize data with your central database management system
- An extremely small memory footprint
- C/C++ or Java development

Full-Featured SQL

UltraLite allows applications on small devices to use full-featured SQL to accomplish data storage, retrieval, and manipulation. UltraLite supports referential integrity, transaction processing, and multi-table joins of all varieties. In fact, UltraLite supports most of the same data types, runtime functions, and SQL data manipulation features as Sybase ASA.

RDBMS Synchronization

UltraLite uses MobiLink synchronization technology to synchronize with industry-standard database management systems. MobiLink synchronization works with ODBC-compliant data sources such as Sybase Adaptive Server Anywhere, Sybase Adaptive Server Enterprise, IBM DB2, Microsoft SQL Server, and Oracle.

You can also use ActiveSync to synchronize your UltraLite applications on Windows CE devices.

Small Footprint

UltraLite provides an ultra-small footprint by generating a custom database engine for your application. This custom engine includes only the features required by your application.

Each UltraLite application has its own database, modeled after a reference Adaptive Server Anywhere database.

C/C++ UltraLite custom database engines for your application can be as small as 50Kb, depending on your deployment platform and the number of SQL statements in the application and the SQL features used.

UltraLite target devices might have no hard disk and tend to have relatively slow processors. The UltraLite runtime employs algorithms and data structures that provide high performance and low memory use. Think of this as an in-memory database.

Supported Development Models

You can develop UltraLite applications in the following ways:

- **C/C++ using the UltraLite C++ API.** The C++ interface exposes tables and SQL statements as objects, with methods to move through the rows of the table or a result set and to execute statements.
- **C/C++ using embedded SQL.** Embedded SQL is an easy way of including SQL statements directly in C or C++ source code.
- **Java using JDBC.** UltraLite Java applications support standard JDBC data access methods.

The UltraLite development tools supplement a supported C/C++ or Java development tool. They manage the generation of the data access code for your application.

UltraLite Generator

During UltraLite application development, you create an Adaptive Server Anywhere *reference database*, which is a model of your UltraLite database. You use the UltraLite generator, which uses the reference database to create the data access and data management code for your application.

The UltraLite generator is a command-line executable with the following syntax:

ulgen -c "connection-string" options, where *options* depend on the specifics of your project.

The UltraLite generator command line customizes its behavior. The following command-line switches are used across development models:

-c. You must supply a connection string, to connect to the reference database; e.g., "DBN=databasename;UID=DBA;PWD=SQL".

-f. Specify the output filename to contain the code.

-j. Specify the UltraLite project name as defined in Sybase Central.

MobiLink

The third part of Sybase's mobile strategy is MobiLink. This step is the middleware that connects the enterprise or desktop database and application with the small device application (and database per se). This is the glue that holds both ends together throughout the entire architecture.

Overview

Many mobile and embedded computing applications are integrated into an information infrastructure. They require data to be uploaded to a *central database*, which consolidates all the data throughout the MobiLink installation, and downloaded from a consolidated database. This bi-directional sharing of information is called *synchronization*.

MobiLink synchronization technology, included in SQL Anywhere Studio along with UltraLite, is designed to work with industry-standard SQL database management systems from Sybase and other vendors. The UltraLite runtime automatically keeps track of changes made to the UltraLite database between each synchronization with the consolidated database. When the UltraLite database is synchronized, all changes since the previous synchronization are uploaded for synchronization.

Consolidated Database

What exactly is the consolidated database? Applications (remote applications or databases such as the UltraLite database definition) synchronize with a central consolidated database. This database is the master repository of information in the synchronization system.

You can build a consolidated database using any supported ODBC-compliant product. You can use a Sybase product, such as Adaptive Server Anywhere or Adaptive Server Enterprise, or you can use a product sold by other companies, such as Oracle, IBM DB2, or Microsoft SQL Server. The consolidated database is not on the small device but on the enterprise or desktop.

Numerous strategies exist for structuring the relations between consolidated and remote database(s). Following are two examples.

You can make the schema of the remote databases a *subset* of the schema of the consolidated database. For example, a table EMP might be repeated among a number of different remote sites, and the consolidated database might use data from column EMP.SALARY in a table called EXPENSE. In this instance, the schemas (table definitions or Data Definition Language) of the consolidated and remote databases are different, although the data is identical and shared. Basically, the table definitions between the consolidated (master) database and the remote database(s) are different in name, but the column definition (that is, integer) and data are the same.

You can also make the schema of the remote database *parallel* in structure to the schema of the consolidated database. Here, the schema of the consolidated database is a reference for the remote database. In the consolidated database, you may already have tables that correspond to each of the remote tables. In this instance, the schemas in the consolidated and remote databases are virtually the same, and the data in the remote is only a subset of the data on the consolidated database.

You write *synchronization scripts* for each table in the remote database and you save these scripts on the consolidated database. These scripts, from their central location on the consolidated database, direct the MobiLink synchronization server in moving data between remote and consolidated databases. One script for a particular remote table tells the synchronization server where to store data uploaded from that remote table in the consolidated database. Another script tells the synchronization server which data to download to the same remote table.

To be a bit more explicit and drive the concept home, the desktop or enterprise holds, in our case, the ASA database. We build a reference database in ASA as the UltraLite reference and create scripts to synchronize the data between the two in MobiLink. These scripts are kept in the consolidated database, which also contains all the data. When the application is compiled, the executable (whether C/C++ or Java) is created along with the specific SQL statements to access the UltraLite database. Then the executable is sent to the small device. The first time the executable is run, the UltraLite database definition is created. Then we synchronize via MobiLink synchronization server to populate the small device UltraLite database with data from the reference database in ASA, which is in effect a subset of the central database also in our case an ASA database. This synchronization uses those scripts kept in the consolidated database specifically for synchronization and not the specific application scripts used in the UltraLite database.

A reference database is an ASA database used in the development of UltraLite clients, or as a convenience in the creation of remote Adaptive Server Anywhere clients. You can use a single ASA database as both a reference and consolidated database during development. Databases made with other products cannot be used as reference databases.

Central Database Subset

Mobile and embedded databases might not contain all the data that exists in the consolidated database.

The tables in each UltraLite database can have a subset of the rows and columns in the central database (consolidated database). For example, a customer table might contain over 100 columns and 100,000 rows in the consolidated database, but the UltraLite database might only require 4 columns and 1000 rows. MobiLink allows you to define the exact subset to be downloaded to each remote database.

Flexibility

MobiLink synchronization is flexible. You define the subset of data using the native SQL dialect of the consolidated database management system. Tables in the UltraLite database can correspond to tables in the consolidated database, but you can also populate an UltraLite table from a consolidated table with a different name, or from a join of one or more tables.

Conflict Resolution

Mobile and embedded databases frequently share common data. They also must allow updates to the same data. When two or more remote databases simultaneously update the same row, the conflict cannot be prevented. It must be detected and resolved when the changes are uploaded to the central database. MobiLink synchronization automatically detects these conflicts. The conflict resolution logic is defined in the native SQL dialect of the central DBMS.

MobiLink Synchronization Server

An UltraLite application synchronizes with a central consolidated database through the *MobiLink synchronization server*. This server provides an interface between the UltraLite application and the database server.

You control the synchronization process using *synchronization scripts*. These scripts might be SQL statements or procedures written in the native language of the consolidated DBMS, or they might be Java classes. For example, you can use a SELECT statement to identify the columns and tables in the consolidated database that correspond to each column of a row to be downloaded to a table in your UltraLite application. Synchronization designs can specify mappings between tables and rows in the remote database with tables and rows in the consolidated database. The only restriction is that columns match across both databases.

You direct the actions of the MobiLink synchronization server by writing scripts. These scripts are stored in tables that must reside in the consolidated database. An additional table contains information about each remote user. These tables have the same structure in non-Sybase database management systems, but the columns are of equivalent, native types.

Each script controls a particular event during the synchronization process.

The MobiLink synchronization server can handle multiple simultaneous synchronization requests. It does so by maintaining a pool of connections with the database server and temporarily assigning these as required when remote devices connect and synchronize.

The synchronization definition is a version 7.0 database object describing data in an Adaptive Server Anywhere remote database that is to be synchronized with a particular MobiLink synchronization server. When using Adaptive Server Anywhere 8.0 or later, publications and synchronization subscriptions should be used instead.

Synchronization Streams

Synchronization occurs through a synchronization *stream*. Supported streams include TCP/IP, HTTP, HotSync, Scout Sync, ActiveSync, and direct serial port communication. Regardless of the stream, you control the synchronization process using the same SQL scripts defined in your consolidated database.

MobiLink Synchronization Process

Synchronization technologies aim for the timely transfer of data across potentially numerous physically distinct databases.

Synchronization transfers database rows between a consolidated and remote database. Table A at the consolidated site has different data than its counterpart Table B at a remote site. This section describes one way in which MobiLink synchronization can process a simple synchronization task and transfer needed data.

Synchronization begins when a MobiLink remote site opens a connection to a MobiLink synchronization server. During synchronization, a MobiLink client at the remote site uploads database changes made to the remote database since the previous synchronization. On receiving this data, the MobiLink synchronization server updates the consolidated database, and then sends back all relevant changes to the remote site. In Figure 10.5, two inserts have taken place, one in each database.

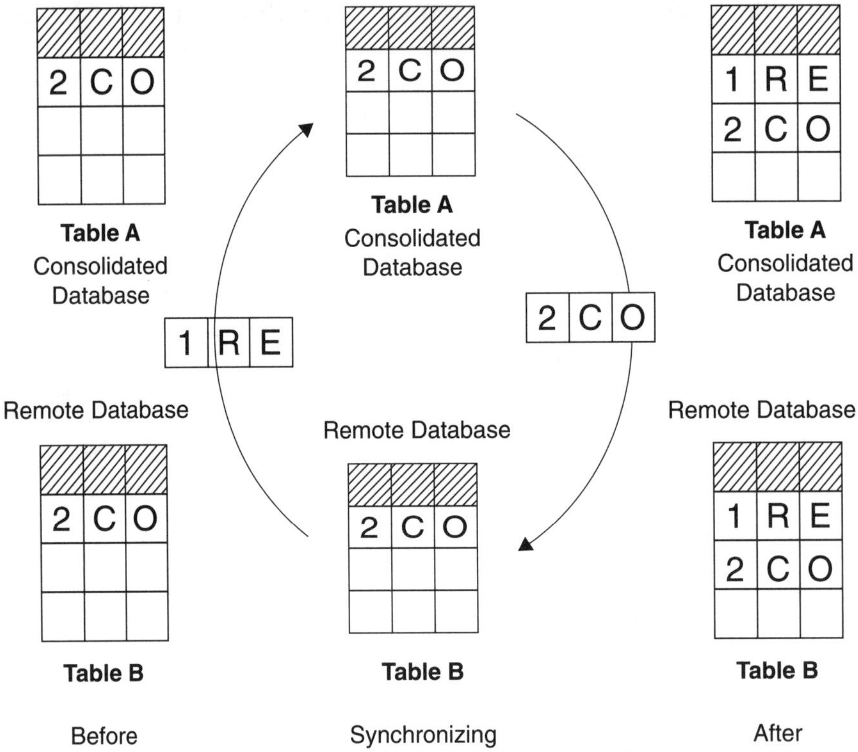

Figure 10.5 Synchronization process.

Record 1RE in the remote database is uploaded to the consolidated database, and Record 2CO in the consolidated database is downloaded to the remote database. The remote site incorporates the entire set of changes, and then optionally sends back a confirmation and closes the connection. At the end of this synchronization session, the databases are consistent.

The After view in Figure 10.5 shows the tables in each database with consistent data after the synchronization process.

Mobile Synchronization System

MobiLink is a session-based synchronization system that allows two-way synchronization between many remote databases and a single ODBC-compliant consolidated database. The consolidated database holds the master copy of all the data. Remote databases can be either Adaptive Server Anywhere or UltraLite databases.

The MobiLink synchronization server acts as a synchronization manager between client applications and the consolidated database.

In addition to managing data flow between consolidated and remote databases, the MobiLink synchronization server carries out coordination, automation, communications, monitoring and reporting, performance improvement, and security.

The System Pieces (according to Sybase)

The following are some of the major components of the synchronization system:

Consolidated database. The central repository of information. This database contains the master copy of all information in the replication system.

Consolidated database server. The server, or DBMS, that manages the consolidated database. This server can be a Sybase product, such as Adaptive Server Anywhere or Adaptive Server Enterprise, or it might be a supported system made by another company.

ODBC connection. All communication between the MobiLink synchronization server and the consolidated database occurs through an ODBC connection. ODBC allows the synchronization server to utilize a variety of consolidated database systems.

MobiLink synchronization server. This server manages the synchronization process and provides the interface between all MobiLink clients and the consolidated database server.

Network. The connection between the MobiLink synchronization server, dbmlsrv8, and the MobiLink client utility, dbmlsync, can use a number of protocols.

MobiLink client. Two types of clients are supported: UltraLite applications and Adaptive Server Anywhere databases. Either or both can be used in a single MobiLink installation.

MobiLink Features (according to Sybase)

The MobiLink synchronization server is adaptable and flexible. MobiLink behavior can be adjusted using a comprehensive range of command-line switches for the MobiLink synchronization server, *dbmlsrv8*, and the Adaptive Server Anywhere synchronization client, *dbmlsync*. You can set a number of switches on the typical MobiLink server or client command line to manage the following:

Data coordination. MobiLink synchronization coordinates data transfer. It allows you to choose selected portions of the data for synchronization.

MobiLink synchronization also allows you to resolve conflicts between changes made in different databases. The synchronization process is controlled by synchronization logic, which can be written in SQL or Java. Each piece of logic is called a *script*.

Automation. MobiLink can perform synchronization automatically. MobiLink has a number of automated capabilities. The MobiLink synchronization server can be instructed to generate scripts suitable for snapshot synchronization, or instructed to generate example synchronization scripts. It can also automatically add users for authentication.

Communications. MobiLink synchronization can occur through numerous communication streams, including TCP/IP, HTTP, and direct serial port communication.

Monitoring and reporting. MobiLink offers you the capability to monitor synchronization. You can monitor scripts, schema contents, row-count values, script names, translated script contents, and row values.

Performance improvement. A number of ways exist to tune MobiLink performance. You can adjust the degree of contention, upload cache size, number of database connections, number of worker threads, logging verbosity, or BLOB cache size.

MobiLink Synchronization Characteristics (according to Sybase)

MobiLink synchronization is characterized by a number of attributes:

Two-way synchronization. Changes to a database can be made at any location.

Choice of communication streams. Synchronization can be carried out over TCP/IP, HTTP, or serial links. Palm devices can synchronize through HotSync. Windows CE devices can synchronize using ActiveSync.

Remote initiated. Synchronization between a remote database and a consolidated database is initiated at the remote database.

Session-based synchronization. All changes are uploaded in a single transaction and downloaded in a single transaction. At the end of each successful synchronization, the consolidated and remote databases are consistent.

Transactional integrity. MobiLink ensures transactional integrity: Either a whole transaction is synchronized, or none of it is synchronized. This ensures transactional integrity at each database.

Data consistency. MobiLink operates using a *loose consistency* policy: All changes are synchronized with each site over time in a consistent manner, but different sites might have different copies of data at any instant.

Adaptability. MobiLink can adapt to popular consolidated databases. MobiLink can communicate with many commercially available database servers, including Oracle, Microsoft SQL Server, Adaptive Server Enterprise, IBM DB2, and Adaptive Server Anywhere.

Wide variety of hardware and software platforms. As listed previously, a variety of widely used database management systems can be used as a MobiLink consolidated database. Remote databases can be Adaptive Server Anywhere or UltraLite databases. MobiLink synchronization server runs on Windows or Unix platforms. Adaptive Server Anywhere runs on Windows, Windows CE, or Unix machines. UltraLite runs on Palm, Windows CE, VxWorks, or Java-based devices.

Flexibility. The MobiLink synchronization server uses SQL or Java scripts to control the upload and download of data. The scripts are executed according to an event model during each synchronization. Event-based scripting provides great flexibility in the design of the synchronization process, including such features as conflict resolution, error reporting, and user authentication.

Scalability and performance. The MobiLink synchronization server is multithreaded, and multiple MobiLink servers can be run simultaneously using load balancing. MobiLink provides extensive monitoring and reporting facilities.

Easy to get started. Simple MobiLink installations can be constructed quickly. More complex refinements can be added incrementally for full-scale production work.

MobiLink Quick Start

The first step is to set up your consolidated and remote databases for MobiLink synchronization. Setting up MobiLink system tables properly on your consolidated database allows you to write synchronization scripts for your application.

The general procedure for setting up your database for use with MobiLink is as follows. This description applies to Adaptive Server Anywhere remote databases.

To set up your databases for MobiLink synchronization:

1. Create a consolidated database and schema.
2. Create a remote Adaptive Server Anywhere database and schema.
3. Create a DSN for your consolidated database.
4. Start the consolidated database.

5. Run one of the setup scripts, appropriate for your consolidated database, located in %ASANY8%\MobiLink\setup\, using an administrative or Interactive SQL tool for your database server (see Table 10.1).

The next procedure is to identify the tables you want synchronized.

To identify the tables and columns you want synchronized:

1. Start Sybase Central, and connect to the consolidated DSN from the MobiLink Synchronization plug-in.
2. Add the tables you want to synchronize into the synchronized tables folder. Do not close down the consolidated database, as you will need to stay connected for the next procedure.

Next, you make a synchronization subscription and publication on the remote database.

To make a synchronization publication and subscription on the remote database:

1. Using Sybase Central, connect to your remote database.
2. Create a publication.
3. Create a synchronization user.
4. Create a synchronization subscription for your user.

The next setup procedure is to write scripts to upload and download data.

To write scripts to upload and download data from each table:

1. Start the MobiLink synchronization server, dbmlsrv8, on the consolidated database using the -za switch.
2. Run the MobiLink client, dbmlsync, on the remote database. Use, for example, the following switches: dbmlsync -c ... -e SendColumnNames='on'.

Table 10.1 Database Setup Scripts

CONSOLIDATED DATABASE	SETUP SQL FILE
Oracle	syncora.sql
IBM DB2	syncdb2long.sql (also syncdb2.sql)
Microsoft SQL Server	syncmss.sql
Adaptive Server Enterprise	syncase.sql

These are the basic steps required to set up your consolidated and remote databases for MobiLink synchronization.

Vineyards Application

Let's look at the application. We'll use an ASA database on Windows 2000, build the application with AppForge, and run it on a Compaq iPAQ. The beauty of AppForge is that a Visual Basic programmer can easily build the application and AppForge will convert all to the required C language for the device.

We'll tackle two things with this application. First, it will show the use of ASA, and second, it will be the extension of a data warehouse. Since manipulating data on the mobile unit is difficult, the trick for data warehousing is to pre-aggregate the required table just as we've done with SalesHistoryCount. The SalesHistory table has multiple rows, one for each month per customer and specific wine. The SalesHistoryCount table tallies all months in one table per customer and specific wine. This method allows for a simpler manipulation of the data once on the wireless device.

This application was created for a fictional wine salesperson for a vineyard called Vineyards (our imagination was overflowing when we thought of that name :-). Sure, it could be enhanced with dozens of features (the application, not our imagination), but we think that what we've created will do the trick for now.

Flow Chart

It starts with an introduction page, which jumps to a main menu page. From this area, the user can view all Customers, Wines, and Sales History information, and list and/or place orders. Of course, there is the customary About page and the usual navigation. To get an understanding of the overall application, we've included the following flowchart in Figure 10.6 along with the database and screens. Later, we'll explain the application and its pros and cons in detail. The Sybase in this application is the ASA database on Windows 2000.

As you can see, the application is not just one program, but a mixture of synchronized programs, each specific to a unique task, as we'll explain later.

ASA Database and Tables

As mentioned earlier, to create a database in Adaptive Server Anywhere, we can use the command line or Sybase Central. To invoke the tool, simply click on >Start >Programs >Sybase SQL Anywhere 8 >Sybase Central.

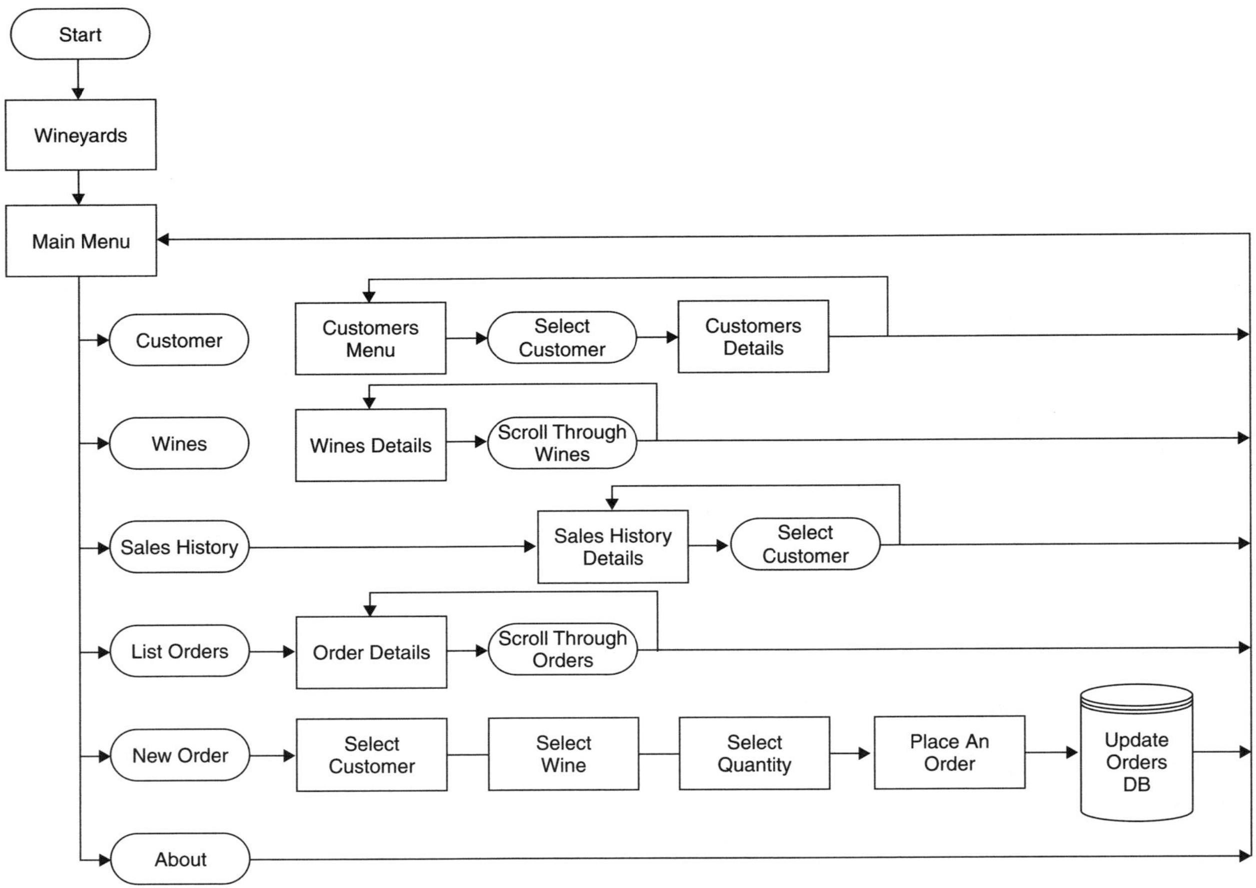

Figure 10.6 Vineyard application flow.

The first thing to do is to create a database by double-clicking on the first item on the right portion of the screen after clicking on Utilities in the left portion of the screen. Click Next on the screen that appears, and then select the *Create a database on this machine* option. In our case, we want to create the ASA database on our desktop computer rather than on a remote server.

Following the Next button will bring you to the screen where you must specify the directory and choose a filename. This filename will be your physical database name and its location on the desktop.

Here the database name is Vineyards and is physically located in the D root directory. Don't worry about a file extension, as the wizard will add the .db extension later. Click Next to continue. Do not select the *Create this database for Windows CE* unless you want to use embedded SQL within an application on the mobile device, which we don't want to do in our example.

Next, the log file for the database will be automatically set using the database name and a .log extension. Just make sure the file is in a directory of your choice. We usually put the database log in the same directory as the physical database to ensure that all database components are physically located in the same directory.

Click Next again, and once again to bypass the mirror log file creation. We won't get into this feature, as it's a bit more advanced than is required right now. Then accept the defaults by leaving the *Install jConnect meta-information support* and click Next again, and then once more to bring you to the page size portion of the wizard.

We highly suggest keeping the recommended default page size of 2K. If not, you should really have a good reason, which is usually a calculated effort to ensure efficient buffer size or maybe extra large table rows. Either way, most systems including networks are optimized for 2K block sizes, so we highly recommend keeping the default settings.

Click Next again, then keep the default collation, and click again onto the next to last page. You could at this point check the Connect to the new database box, which would bring you directly into the database, but instead let's look at another option. The reason is, if you exit Sybase Central, how will you reconnect to the database? Let's look at this alternative method, but first, click on Finish to create the database. A screen will pop up showing what's being done in the creation process and finally will disappear when the database is created.

Okay, so the database is created, but where is it and how do we connect to it? Click on the Tools option on the main tool bar in Sybase Central and then on the Connection Profiles option. This brings you to an option whereby you can create a permanent connect profile. Click on the new connection at a later time and the database is instantly available. Let's see how this is done.

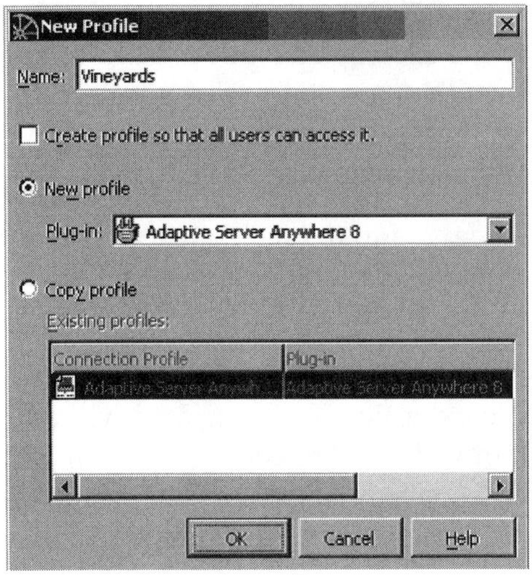

Figure 10.7 New Profile.

Once in the Connection Profile screen, click on New. A second screen will pop up, as shown in Figure 10.7, and then simply fill in a name you'd like to use as the connection profile name, and continue on by clicking OK. Don't confuse the connection profile name with the database name, as they really are different. In our examples, we use the same name, Vineyards, for ease of reference.

Next, enter a username and password under the Identification tab. The default username is DBA in uppercase, and SQL for the password. Then, enter the physical location of the database in the database files area under the Database tab, and click OK. A new connection profile will then be created. The next time you access Sybase Central, click on Tools, Connection Profiles, and then on this new profile you just created. The database will instantly pop up in Sybase Central.

ASA builds a server and a database at the same time—different from ASE, which has its own server and its own database. Expand the server (see Figure 10.8) and you'll see the database with all its objects. The ASA database and server are actually the same as described earlier in the Adaptive Server Anywhere section.

When the connection profile was clicked, something extra happened—an ASA service began (see Figure 10.4). You can see it on your pop-up task (Start) bar next to the time. If you click on this you'll see everything that happened to start the service. It shows the database name, physical location, page size, log name, time the database was started, and so on.

Here's a tech tip: When viewing the database server, do not click on the X button, as this will shut down the database. The trick is to minimize the window, but don't worry—it won't clutter up your task bar.

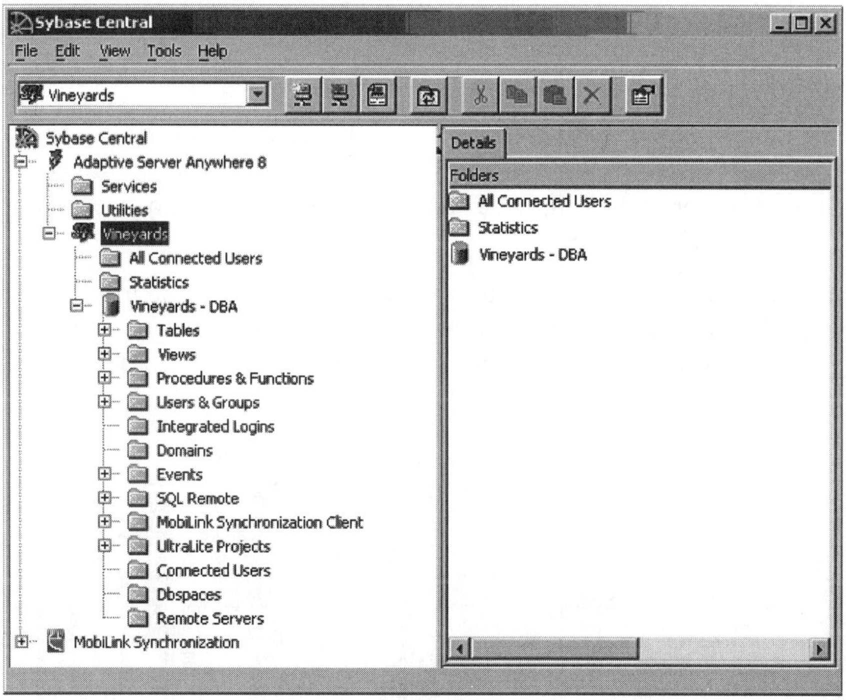

Figure 10.8 Vineyards server and database.

There we have it—a database, an ongoing connection profile, and a started service (server/database) (see Figure 10.9).

Now let's create several tables and populate them with data. Click on the Tables folder in the Vineyards—DBA database. Choose Add Table and enter

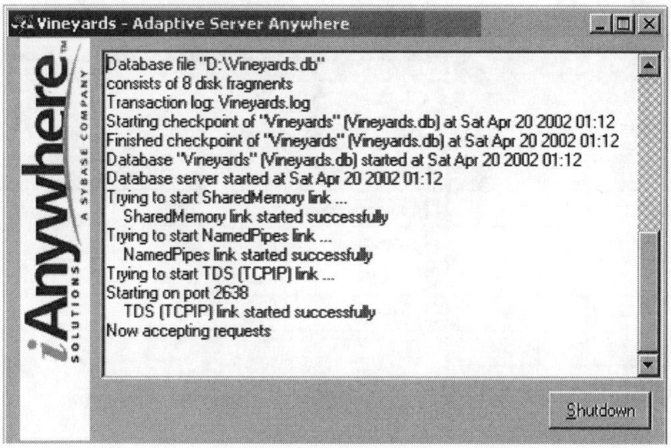

Figure 10.9 Starting Vineyards database.

the tables to be used in the application. Figure 10.10 shows all the tables with all their columns and attributes.

For the sake of simplicity, we'll import predefined data from text files (see Figure 10.11) into our tables. To do this, right-click on each table and choose the *View Data in Interactive SQL*, as shown in the database administration tools section earlier in the chapter. Then, under the Data option, choose Import and follow the wizard steps.

If your data in the text files and the table definitions are in sync, your database will be correctly loaded; otherwise, the import utility will return an error message describing the problem. Note that if you have single quotes in any

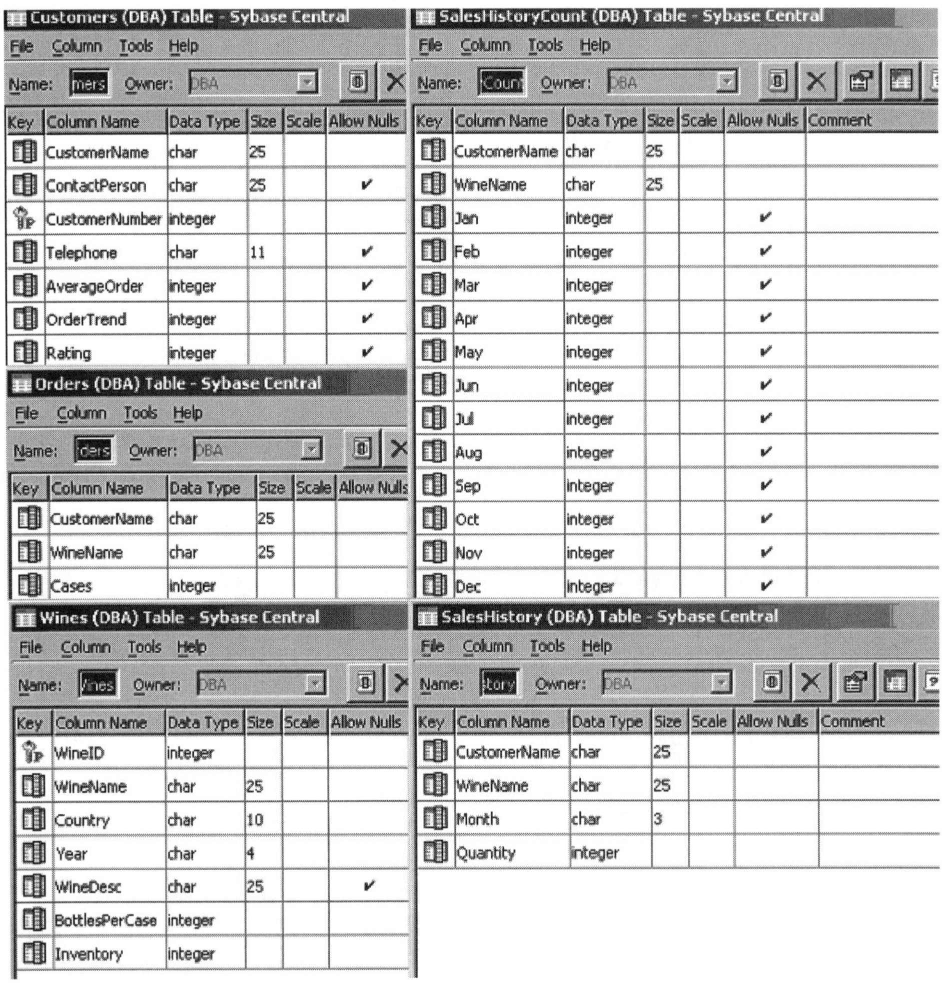

Customers (DBA) Table - Sybase Central

File Column Tools Help

Name: mers Owner: DBA

Key	Column Name	Data Type	Size	Scale	Allow Nulls
	CustomerName	char	25		
	ContactPerson	char	25		✔
	CustomerNumber	integer			
	Telephone	char	11		✔
	AverageOrder	integer			✔
	OrderTrend	integer			✔
	Rating	integer			✔

Orders (DBA) Table - Sybase Central

File Column Tools Help

Name: ders Owner: DBA

Key	Column Name	Data Type	Size	Scale	Allow Nulls
	CustomerName	char	25		
	WineName	char	25		
	Cases	integer			

Wines (DBA) Table - Sybase Central

File Column Tools Help

Name: ines Owner: DBA

Key	Column Name	Data Type	Size	Scale	Allow Nulls
	WineID	integer			
	WineName	char	25		
	Country	char	10		
	Year	char	4		
	WineDesc	char	25		✔
	BottlesPerCase	integer			
	Inventory	integer			

SalesHistoryCount (DBA) Table - Sybase Central

File Column Tools Help

Name: Coun Owner: DBA

Key	Column Name	Data Type	Size	Scale	Allow Nulls	Comment
	CustomerName	char	25			
	WineName	char	25			
	Jan	integer			✔	
	Feb	integer			✔	
	Mar	integer			✔	
	Apr	integer			✔	
	May	integer			✔	
	Jun	integer			✔	
	Jul	integer			✔	
	Aug	integer			✔	
	Sep	integer			✔	
	Oct	integer			✔	
	Nov	integer			✔	
	Dec	integer			✔	

SalesHistory (DBA) Table - Sybase Central

File Column Tools Help

Name: tory Owner: DBA

Key	Column Name	Data Type	Size	Scale	Allow Nulls	Comment
	CustomerName	char	25			
	WineName	char	25			
	Month	char	3			
	Quantity	integer				

Figure 10.10 Vineyards tables.

Figure 10.11 Customer import text file.

char (text) field, simply add another single quote for the import wizard to correctly recognize it, as seen in Figure 10.11. Notice the first entry is *Srdjan's vines*, but is represented as *Srdjan''s vines* to specifically ensure that the single quote is recognized.

Now we're ready for the application, but first we need to connect the data and structure on the desktop database with data and structure on the wireless device. This is done via the conduit.

AppForge Conduit

This product was explained in detail earlier in the book, so we won't go into the wireless device Booster portion at this point. However, as a brief recap of Conduits: The AppForge Conduit will build the connection between the ASA database and the final database files that will reside on the Compaq iPAQ device. The wizard creates one PDB file per ASA database table and also copies the ASA table data. It also creates BAS files for the software to simulate the wireless device on the desktop, which is a really fantastic feature.

To invoke the AppForge Universal Conduit Wizard, begin by clicking >Start > Programs >AppForge, and then click on the Conduit Wizard. Click the Next button to begin. Enter a descriptor name, such as Vineyards for our application, and even though we are programming for a Pocket PC device, enter the Palm Creator ID received from the Palm Web site as shown in Figure 10.12.

The next step is to connect to the ASA database. We set up an ODBC Data Source Name (DSN) and select the entry via the selection window.

The third step is to select the tables to be synchronized. The ODBC DSN connection on the desktop to our Vineyards ASA database will show all available tables. We'll select only four, since these are the only tables we'll be using in the application (see Figure 10.13).

For each table, the synchronization type must be defined. AppForge software was initially used for Palm connectivity, and hence the Palm terms used throughout the wizards. However, rest assured the connectivity for Pocket PCs will be resolved down the line as we progress through the tool's functionality.

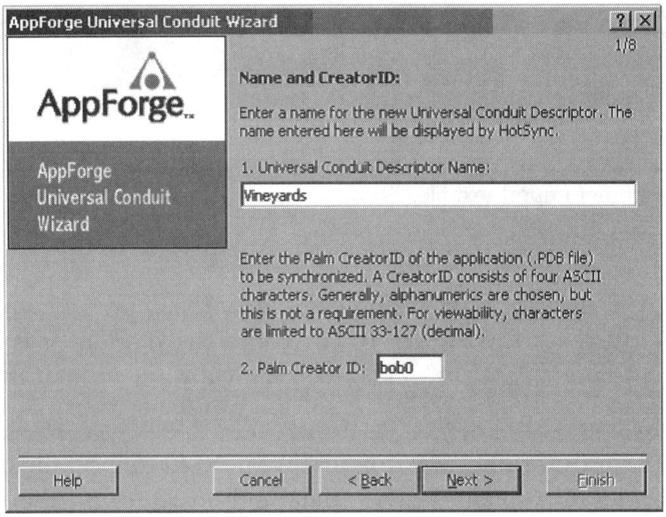

Figure 10.12 Descriptor name.

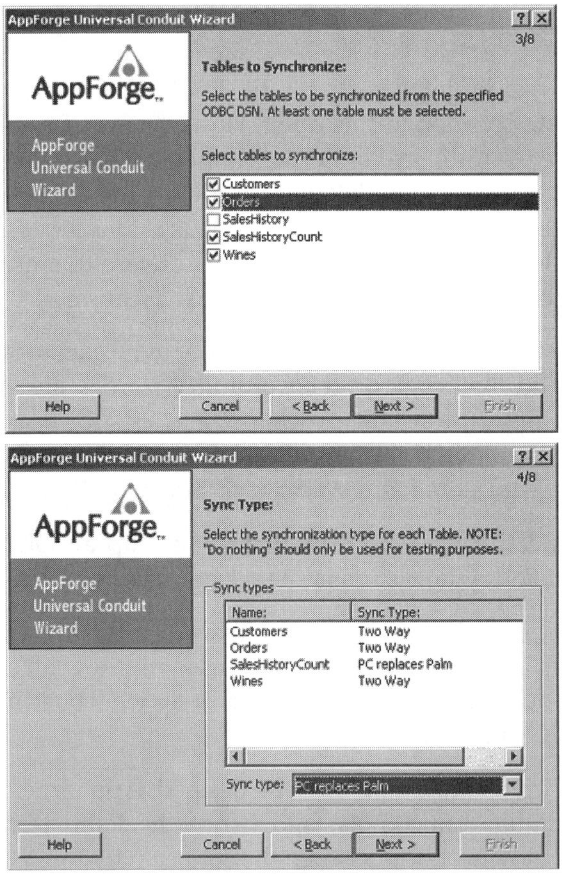

Figure 10.13 Tables and sync types.

As you can see, the application uses two-way synchronization for three tables and only one way for the SalesHistoryCount table. The reason is because this table is a simulation of an extension to a data warehouse and data will only be sent to the Pocket PC and never vice versa.

The fifth of eight steps is to select which table keys to synchronize (see Figure 10.14). This is important because each table row, if changed, must somehow be recognized from its previous entry, and the only method to do this is via a static key field between the desktop and the wireless device. The Sync Keys picture shows the Customers table having a CustomerNumber field, which is the primary key (as seen from the ASA database table creation section earlier). This field will be the Sync Key for this table in our application. Each of the other tables, excluding SalesHistoryCount, will have their own Sync Keys, being their respective primary keys.

The screen shown in Figure 10.15, step six, allows the creation of the PDB files and the BAS files. The PDB files are used on the Pocket PC device and will be transferred to the device later. The BAS files contain the code to access these files, also seen momentarily.

Step seven is shown in Figure 10.16. This step allows you to review all the specifics for the Conduit. Scroll through the settings and review each to ensure that you've selected all the correct attributes. Then, press the Finish button to compile and Upload the Vineyards Conduit.

That's it—the Conduit is built, all files have been created, and the proper PDB database files have been uploaded to the wireless device. The next time you

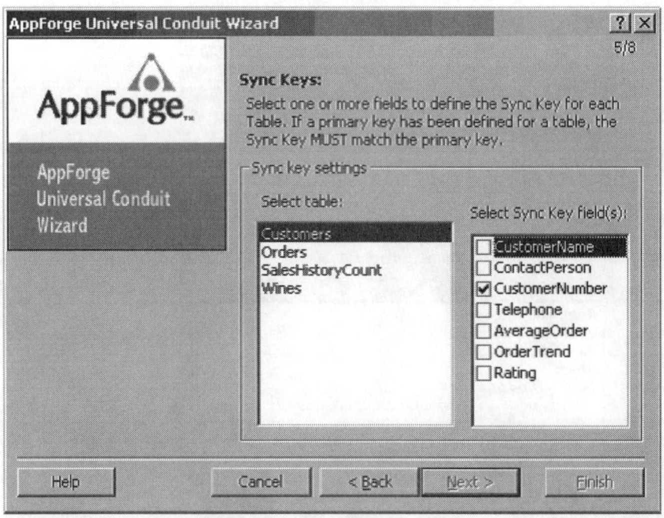

Figure 10.14 Sync Keys.

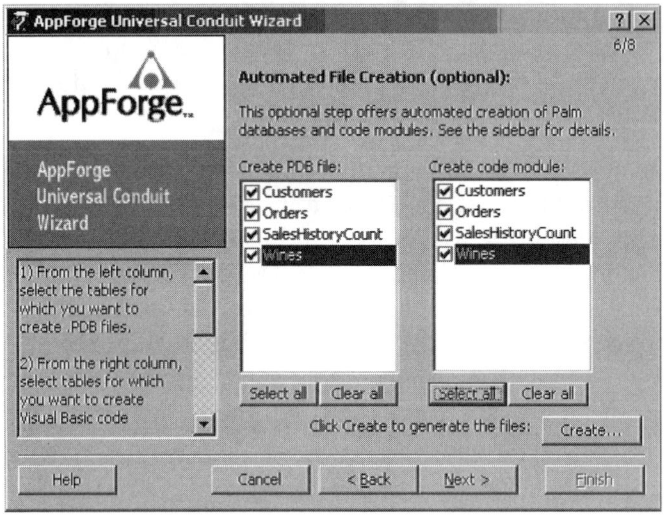

Figure 10.15 PDB file creation.

bring up the AppForge Universal Conduit tool, you'll notice that the new Vineyards Conduit is already available (see Figure 10.17).

If, however, you decide to expand your application by adding another table to the ASA database and it's to be used on the wireless device, as in our Vineyards application, you will have to update the Conduit. From our experience,

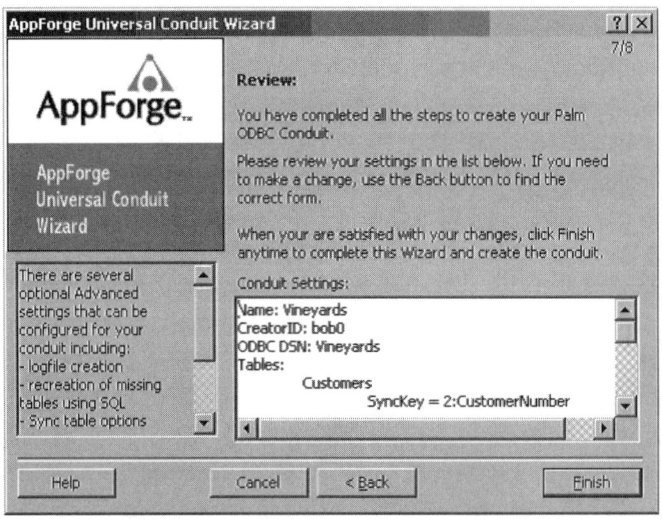

Figure 10.16 Review Conduit settings.

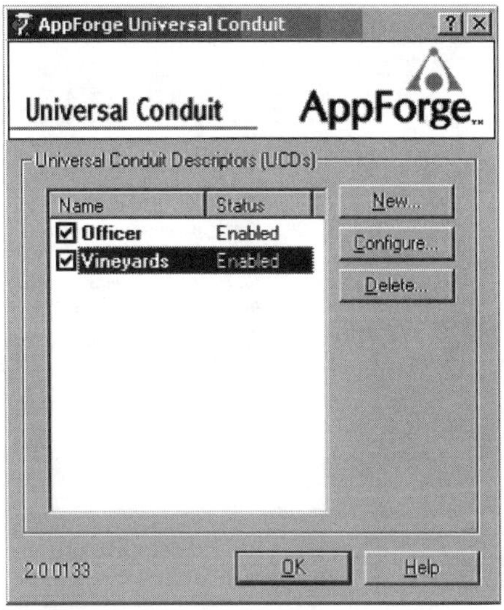

Figure 10.17 AppForge universal conduit descriptors.

the only way to do this is to delete the existing Conduit and recreate it, but this time with the new table. Unfortunately, you can't simply add another table and have the Conduit wizard generate and upload the new .pdb file. Moreover, if you have specific database access code in the BAS file, you will lose it and have to re-add it after you recreate the Conduit again. If you forget the exact syntax of the command, you might spend a little frustrating time recreating it after you figure out why your application won't compile.

You might also have to download the specific PDB file(s) to the device manually. The best way to do this is through Explorer on the desktop, with the Pocket PC device connected to the USB port. That will allow you to check the mobile device files directly. If you use File Explorer on the wireless device, you will not see file extensions, but if the Mobile Device option on the desktop's Explorer is used, all of them will be visible (see Figure 10.18). The best way to transfer files from the desktop to the Pocket PC is through the desktop Explorer. Copy and paste as you would any file to any other directory.

The following is the code for an AppForge-generated BAS module. This one happens to be for the Customers table and is called Customers.BAS. As you can see, the code defines specific routines to be used in the Visual Basic application. For example, in the application simply say *OpenCustomersDatabase*, which translates to quite a number of commands behind the scenes, as seen on lines 40 through 59. The underlying code, shown in Listing 10.1, uses PDB commands to open the database and to test whether all went smoothly.

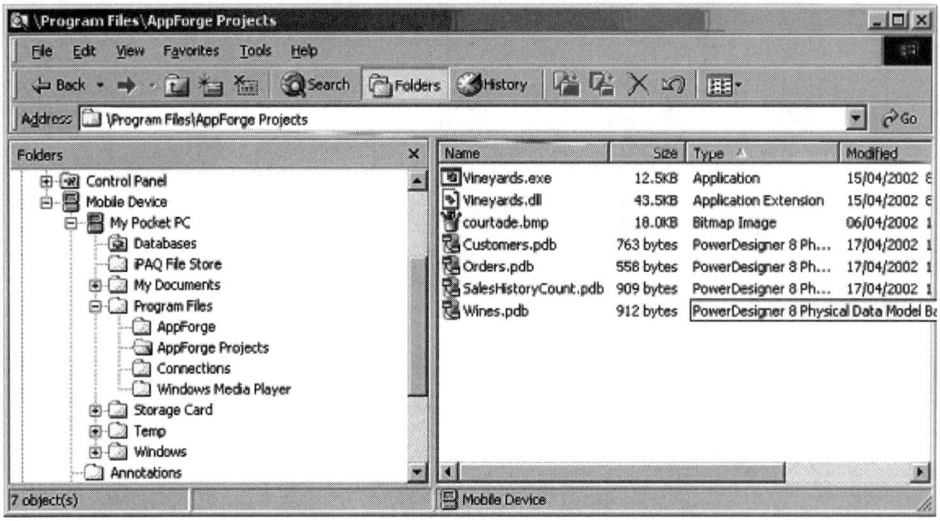

Figure 10.18 Explorer on mobile device.

```
'-------------------------------------------------------------
'      AppForge PDB Converter auto-generated code module
'
'      Source Database:
'      Source Table   : Customers
'
'      Num Records    : 3
'
'      PDB Table Name : Customers
'            CreatorID : bob0
'            TypeID    : DATA
'
'       Converted Time : 07/04/2002 4:01:25 PM
'-------------------------------------------------------------
Option Explicit

' Use these constants for the CreatorID and TypeID
Public Const Customers_CreatorID As Long = &H626F6230
Public Const Customers_TypeID As Long = &H44415441

' Use this global to store the database handle
Public dbCustomers As Long

' Use this enumeration to get access to the converted database
   Fields
Public Enum tCustomersDatabaseFields
CustomerName_Field = 0
```

Listing 10.1 Customers.BAS module.

```
ContactPerson_Field = 1
CustomerNumber_Field = 2
Telephone_Field = 3
AverageOrder_Field = 4
OrderTrend_Field = 5
Rating_Field = 6
End Enum

Public Type tCustomersRecord
CustomerName As String
ContactPerson As String
CustomerNumber As Long
Telephone As String
AverageOrder As Long
OrderTrend As Long
Rating As Long
End Type

Public Function OpenCustomersDatabase() As Boolean
' Open the database
#If APPFORGE Then
dbCustomers = PDBOpen(Byfilename, "Customers", 0, 0, 0, 0,
    afModeReadWrite)
#Else
dbCustomers = PDBOpen(Byfilename, App.Path & "\Customers",
    0, 0, 0, 0, afModeReadWrite)
#End If
If dbCustomers <> 0 Then
'We successfully opened the database
OpenCustomersDatabase = True
Else
'We failed to open the database
OpenCustomersDatabase = False
#If APPFORGE Then
MsgBox "Could not open database - Customers", vbExclamation
#Else
MsgBox "Could not open database - " + App.Path + "\Customers.pdb"
    + vbCrLf + vbCrLf + "Potential causes are:" + vbCrLf + "1.
    Database file does not exist" + vbCrLf + "2. The database path
    in the PDBOpen call is incorrect", vbExclamation
#End If
End If
End Function

Public Sub CloseCustomersDatabase()
' Close the database
PDBClose dbCustomers
dbCustomers = 0
End Sub

Public Function GoToCustomersRecord(MyIndex As Integer) As Boolean
GoToCustomersRecord = PDBGotoIndex(dbCustomers, MyIndex)
End Function
```

Continues

Listing 10.1 *(Continued)*

```
Public Function ReadCustomersRecord(MyRecord As tCustomersRecord)
    As Boolean
ReadCustomersRecord = PDBReadRecord(dbCustomers, VarPtr(MyRecord))
End Function

Public Function WriteCustomersRecord(MyRecord As tCustomersRecord)
    As Boolean
WriteCustomersRecord = PDBWriteRecord(dbCustomers,
    VarPtr(MyRecord))
End Function
```

Listing 10.1 Customers.BAS module. *(Continued)*

The Application

The Vineyard application flows from a single main menu invoked from the introduction page. This introduction page, as shown in Figure 10.20, contains an image. To include this image, we simply used the AppForge AFGraphic icon from the toolbox general menu (see Figure 10.19).

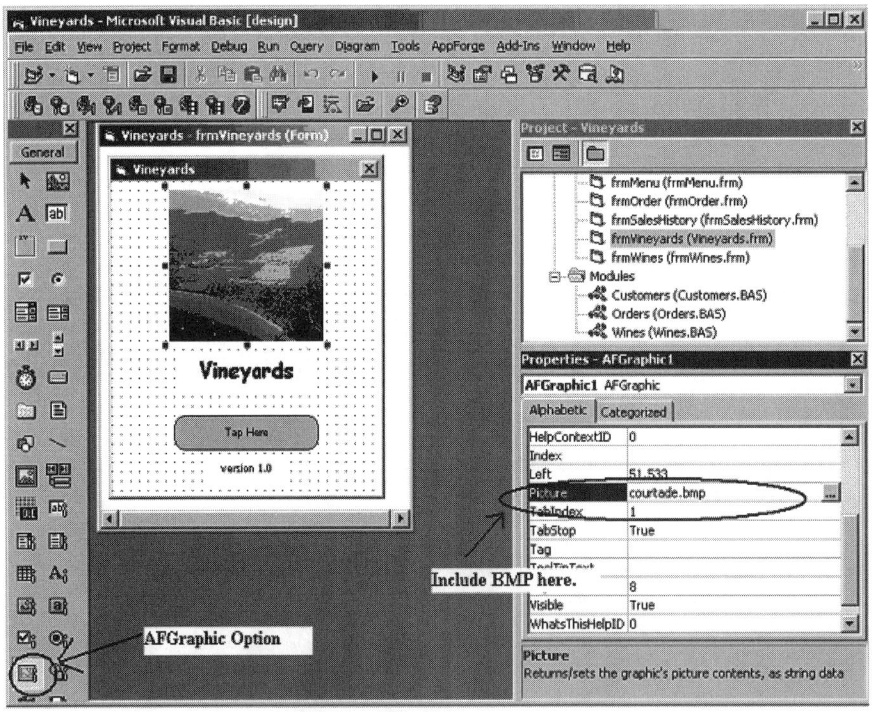

Figure 10.19 Including BMP in application.

We thought this might be of interest just before we introduce the underlying code for the entire application, as it gives a good insight into the AppForge tool and shows the design process in the making. This is a good screen to see the Vineyards application within AppForge and to demonstrate a specific feature such as how to get a picture on a screen. The remainder of this chapter will show the final screens and code.

Since this application, or *project* as AppForge likes to call it, is specific for Pocket PC (as specified when the project was initially created), the screen size is exactly that of the physical device. Therefore, what you see is what you get (well, it is in the actual tool!!).

Let's begin the application from the introduction screen. The first window is the Vineyards application introduction page as seen in the Application Flow diagram earlier (Figure 10.6) and as shown in Figure 10.20.

The behind-the-scenes Visual Basic application code is as follows in Listing 10.2. It's very simple; the "Tap Here" gray button was named TapButton and it has a Click option. The TapButton_Click subroutine process includes hiding the current window, unloading the current window from memory, and showing the next window in the application flow—the Main Menu window called frmMenu. When linking to a new window, we find it important to hide the existing window or else the second window might be superimposed on the first, which ends up looking rather strange. In addition, since the majority of wireless devices have little memory, we find it helpful to remove each window from memory as the application flows for performance reasons.

Figure 10.20 Vineyards window.

```
Option Explicit
Private Sub TapButton_Click()
frmVineyards.Hide
Unload Me
frmMenu.Show
End Sub
```

Listing 10.2 Vineyards code.

Tapping the Tap Here button will flow to the application's main menu as shown in Figure 10.21. This menu contains all the main options of the application. Simply tap on any button to proceed.

There are seven options from which to choose, and therefore seven buttons and seven respective subroutines (Listing 10.3). All the subroutines are basically the same. Each subroutine process is as the introduction page, hide the current page, unload the current page from memory, and then flow to the next appropriate screen. Very simple logic, as it simply links off to another program.

Let's begin with the first submenu, the Customers menu. When the Customers button on the main menu is tapped, flow will be directed to the Customers menu (see Figure 10.22). Once on the page, all customers will be included dynamically in the drop-down list box. All the end user has to do is tap on the list box and select the specific customer desired to continue flow to the next screen, which will show that customer's details, as we'll soon see.

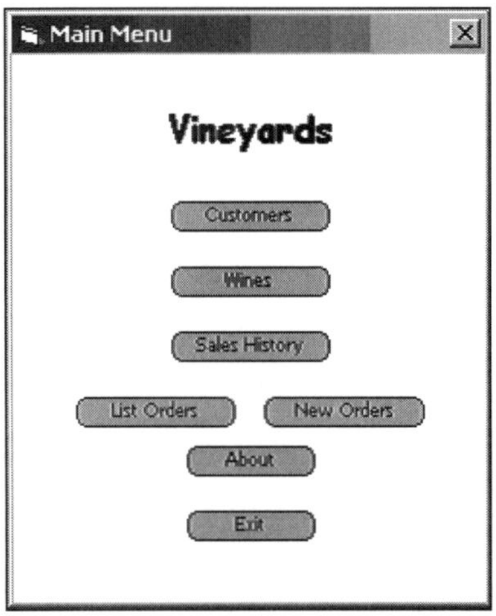

Figure 10.21 Vineyards main menu.

```
1. Option Explicit
2. Private Sub AboutButton_Click()
3. frmMenu.Hide
4. Unload Me
5. frmAbout.Show
6. End Sub

7. Private Sub CustomersButton_Click()
8. frmMenu.Hide
9. Unload Me
10.frmCustomers.Show
11. End Sub

12. Private Sub ListOrderButton_Click()
13. frmMenu.Hide
14. Unload Me
15. frmListOrders.Show
16. End Sub

17. Private Sub SalesHistory_Click()
18. frmMenu.Hide
19. Unload Me
20. frmSalesHistory.Show
21. End Sub

22. Private Sub WinesButton_Click()
23. frmMenu.Hide
24. Unload Me
25. frmWines.Show
26. End Sub

27. Private Sub OrderButton_Click()
28. frmMenu.Hide
29. Unload Me
30. frmOrder.Show
31. End Sub

32. Private Sub ExitButton_Click()
33. frmMenu.Hide
34. Unload Me
35. frmVineyards.Show
36. End Sub
```

Listing 10.3 Vineyards main menu code.

There are three main pieces to this program. First is to define what happens once the Menu button is tapped. Second is the underlying logic for when a customer is selected. A new type of process is used in this program and is invoked when the form (screen) is loaded. Let's start at this third portion, lines 15 through 30, shown in Listing 10.4.

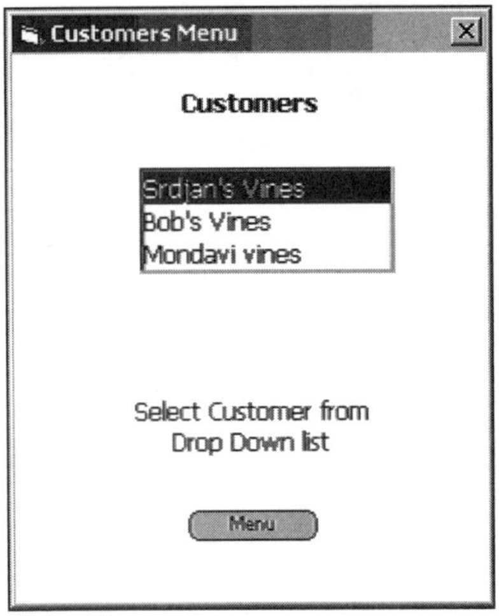

Figure 10.22 Customers menu.

The first thing we must do is to obtain the customer's name from the database (Customers.pdb file). Line 16 shows how the Customers database is opened, as explained earlier. Then, it is a good programming technique to clear the AFComboBox, thus ensuring that nothing is lingering around from any previous routines. Now the fun begins. The idea is to loop through all the records in the Customers database. Therefore, move the cursor to the first customer record, test to ensure that there are records, and if so, continue looping through the set. While doing so, list them in the AFComboBox. Line 19 does the test to see if the end of file was reached; if not, then the process continues. If the end is at hand, the logic will display the *No Customers* message in the AFComboBox. Lines 22 through 26 perform the looping through the Customer records. The loop will continue until the end of file and then end, flowing to line 28 where the database is closed. Line 29 sets the current entry in the drop-down list box to zero, which means the first customer name in the list.

Line 23 reads the customer record at the cursor position, line 24 adds the Customer name to the AFComboBox to appear on the screen, and line 25 moves the cursor to the next entry in the Customer database—and the loop continues again until no more Customer records exist. This routine is what sets the Customer names on the screen when the screen first appears after selection.

The next question is: What happens when a customer name is selected from the AFComboBox? This logic is in the second subroutine called AFComboBox2 _Click. When the user taps on the combo box, a drop-down list of all customers will appear. When a specific customer is selected, this subroutine is invoked (lines 8 through 14). Line 9 opens the database, and line 10 will set the database record currency to a specific entry based on the physical entry in the drop-down list box (starting at zero, not one). All that's left to do, at this point, is hide the current screen, unload the current screen from memory, and flow to the next program/form called frmCustDetails, as shown in line 13. The database will remain open during the transition to the next screen, and processing can continue as if a single program is executing.

The first subroutine is typical, as it performs the usual housekeeping and returns control to the main menu screen via line 6.

The application and database were specifically designed together. Each customer's information was specifically set in one record within the database, so the program could easily read what it requires in one instance and report it on the screen. This method is very efficient and reduces database scan time. This is a typical data warehouse extension for small devices.

```
1. Option Explicit
2. Dim MyRecord As tCustomersRecord

3. Private Sub MenuButton_Click()
4. frmCustomers.Hide
5. Unload Me
6. frmMenu.Show
7. End Sub

8. Private Sub AFComboBox1_Click()
9. OpenCustomersDatabase
10. GoToCustomersRecord AFComboBox1.ListIndex
11. frmCustomers.Hide
12. Unload Me
13. frmCustDetails.Show
14. End Sub

15. Private Sub Form_Load()
16. OpenCustomersDatabase
17. AFComboBox1.Clear
18. PDBMoveFirst (dbCustomers)
19. If PDBEOF(dbCustomers) Then
20. AFComboBox1.AddItem "No Customers"
21. Else
22. While Not PDBEOF(dbCustomers)
23. ReadCustomersRecord MyRecord
24. AFComboBox1.AddItem MyRecord.CustomerName
25. PDBMoveNext (dbCustomers)
26. Wend
27. End If
28. CloseCustomersDatabase
29. AFComboBox1.ListIndex = 0
30. End Sub
```

Listing 10.4 Customers code.

In the legacy system on the desktop or network, the Customer information might be found in multiple tables, and replicating the exact database and data structure would be extremely inefficient. Therefore, we merge whatever customer information the wireless device application requires in one table, and the application can then access it in one simple process.

The final step in the customer process is to view the customer details. This screen (see Figure 10.23) has the customer's name, contact information, telephone number, average order, order trend, and rating. The final three bits are again a classic data warehouse extension whereby information from, presumably, multiple sources is gathered and centralized at the customer level for easy reporting as shown in Figure 10.23.

To select another customer, the user can tap on the Customer Menu button. To return to the application main menu, tap on the Menu button.

Line 2 of Listing 10.5 declares a variable called MyRecord and has the identical characteristics as the tCustomersRecord, which is defined in the Customer.BAS module shown earlier.

Lines 3 through 7 are invoked when the Customer Menu button is pressed, and lines 8 through 12 are executed when the Menu button is tapped.

Lines 13 through 22 are initiated when the screen first appears. Remember the previous Customer Menu—when a customer entry was tapped from the drop-down list box, part of the routine was to open the database and set the database pointer to the proper record as selected by the user. When this screen is initiated, the Form_Load subroutine will read the database record (line 14)

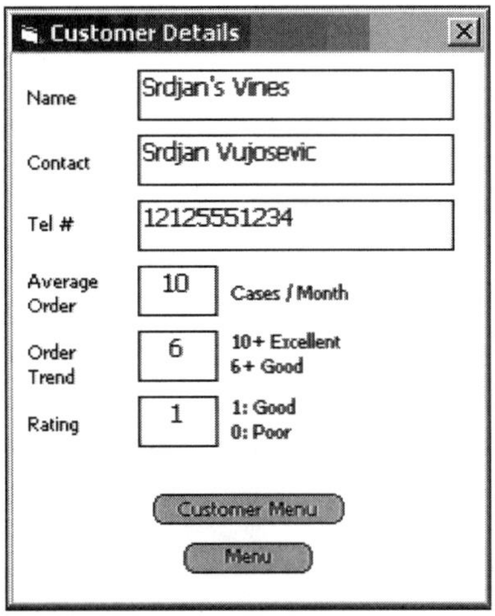

Figure 10.23 Customer details.

and set the screen fields to the database fields as seen on lines 15 through 20. Line 21 closes the database, and with the ending of the subroutine on line 22, the screen is displayed with all the information held in the database for that particular customer.

```
1. Option Explicit
2. Dim MyRecord As tCustomersRecord

3. Private Sub CustMenuButton_Click()
4. frmCustDetails.Hide
5. Unload Me
6. frmCustomers.Show
7. End Sub

8. Private Sub MenuButton_Click()
9. frmCustDetails.Hide
10. Unload Me
11. frmMenu.Show
12. End Sub

13. Private Sub Form_Load()
14. ReadCustomersRecord MyRecord
15. AFTextBox1.Text = MyRecord.CustomerName
16. AFTextBox2.Text = MyRecord.ContactPerson
17. AFTextBox3.Text = MyRecord.Telephone
18. AFTextBox4.Text = MyRecord.AverageOrder
19. AFTextBox5.Text = MyRecord.OrderTrend
20. AFTextBox6.Text = MyRecord.Rating
21. CloseCustomersDatabase
22. End Sub
```

Listing 10.5 Customers Detail code.

Moving on to the next main menu option of Wines. From the main menu, select the Wines button and flow will move to the following screens. We've included the final version you'd see on the screen and the programmer's version you'd see when developing the screen in Figure 10.24. This was done for ease of reference when viewing the code and referencing it back to the screen fields.

The logic behind this screen (Listing 10.6) is a little more advanced. On this screen, the first time in (Form_Load subroutine) will result in the display of the first wine in the wine database. Then, the user can scroll through the wines, one by one, using the arrow keys. The left-pointing arrows are used to move backward, and the right-pointing arrows are used to move forward in the list. This is a very simple list to show the basics of scrolling.

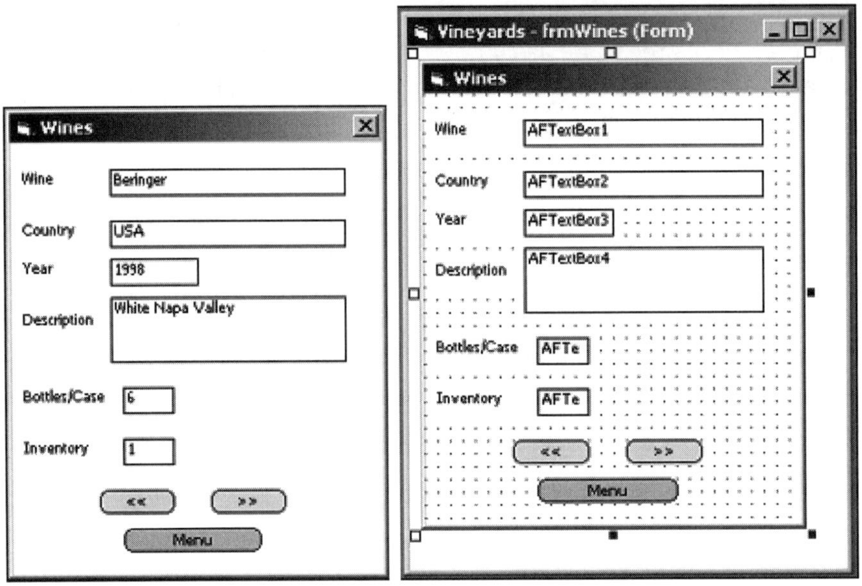

Figure 10.24 Wine menu.

The Form_Load routine is invoked once and only after the user taps the *Wines* button on the main menu. Here, this routine will open the Wines database, sort the wines by name (line 23), set the cursor to the first wine in the database, and call a public subroutine called DisplayData. This subroutine, lines 27 through 44, will display blank and zeros in the fields along with the word *None* in the wine name field if there are no wines in the database. This is determined from line 28, which simply checks the number of records in the database.

If records do exist, the first record is read into the defined variable MyRecord (line 2), and each screen field is loaded with the appropriate data from the record (lines 30 through 35).

To scroll forward, the end user would tap the right pointing arrows button and the NextButton_Click subroutine is invoked. Because the database is still opened, the database cursor can move to the next record in the database using the command on line 10. If the end of file is not reached, the DisplayData routine is called to display the information for the next wine.

If the backward button is tapped, the PreviousButton_Click, line 3, is invoked and the cursor is moved in the opposite direction, retrieving previously viewed information from the database.

The database is finally closed when the Menu button is tapped. Line 16 of the MenuButton_Click subroutine performs this final database housekeeping task before the typical hiding of the current screen, unloading from memory, and linking to the appropriate screen (Main Menu) is called.

```
Option Explicit
Dim MyRecord As tWinesRecord

Private Sub PreviousButton_Click()
PDBMovePrev (dbWines)
If Not PDBBOF(dbWines) Then
DisplayData
End If
End Sub

Private Sub NextButton_Click()
PDBMoveNext (dbWines)
If Not PDBEOF(dbWines) Then
DisplayData
End If
End Sub

Private Sub MenuButton_Click()
CloseWinesDatabase
Me.Hide
Unload Me
frmMenu.Show
End Sub

Private Sub Form_Load()
OpenWinesDatabase
PDBSetSortFields dbWines, tWinesDatabaseFields.WineName_Field
PDBMoveFirst (dbWines)
DisplayData
End Sub

Public Sub DisplayData()
If PDBNumRecords(dbWines) > 0 Then
ReadWinesRecord MyRecord
AFTextBox1.Text = MyRecord.WineName
AFTextBox2.Text = MyRecord.Country
AFTextBox3.Text = MyRecord.Year
AFTextBox4.Text = MyRecord.WineDesc
AFTextBox5.Text = MyRecord.BottlesPerCase
AFTextBox6.Text = MyRecord.Inventory
Else
AFTextBox1.Text = "None"
AFTextBox2.Text = ""
AFTextBox3.Text = ""
AFTextBox4.Text = ""
AFTextBox5.Text = 0
AFTextBox6.Text = 0
End If
End Sub
```

Listing 10.6 Wine menu code.

The next Main Menu option is Sales History. This screen (see Figure 10.25) is similar to the Customer's drop-down list screen except that when the customer is selected, the information appears on the current screen rather than the flow jumping to another screen. The purpose of this screen is to give the salesperson a quick view of all the customers' purchases for a particular wine over the past calendar year.

This screen represents another basic data warehouse extension to the wireless device. Rather than computing the monthly sales for the calendar year on the device, everything is predetermined on the desktop. This means that most processing resources have already been taken before the information is uploaded to the device, which saves quite a bit of time, giving fast response times for the wireless device user.

In the database on the desktop, the SalesHistory table holds the count of sales per wine per customer per month. A routine would be run on this table on the desktop to create an aggregated version resulting in the SalesHistoryCount table. Then, we read the table, same idea done earlier.

This program (Listing 10.7) gets a bit tricky, as each iteration must ensure that the exact same drop-down list is created but the selected customer name is prominently displayed. Let's look at the code starting with the Form_Load subroutine.

Lines 41 through 72 perform the Form_Load routine. First, the SalesHistory-Count database is opened (line 42), the drop-down box is cleared (line 43), and the cursor is set to the first record in the database (line 44).

If records exist in the database, line 45, a loop is invoked (lines 48 through 52). This loop will read the record at the cursor position and add the customer

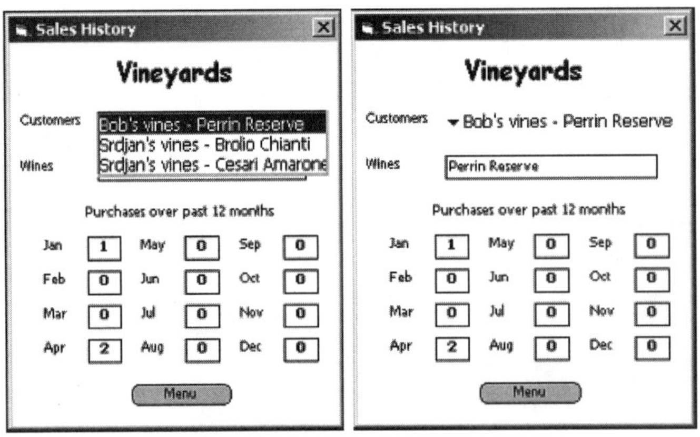

Figure 10.25 Sales History menu.

name plus the wine name to the ComboBox until all the records in the database have been read.

Because this is the first time on the screen, the combobox display is set to the first record in the list. Then, the database pointer is positioned to the first record, it is re-retrieved, and the screen fields are filled with the information from the record. The database is then closed and the screen is displayed.

Other than tapping the Menu button on the screen, the only other application function on this screen is to tap the customer name to view the sales history. When this is done, the CustAFComboBox1_Click subroutine is invoked (lines 9 through 40). Here the program keeps track of the specific combobox item selected by storing it in the CustRow variable on line 10. Then the database is opened (line 11), the combobox is cleared of any previous entries (line 12), and the cursor is set to the first record in the database (line 13). It's important to clear the combobox before we get started, otherwise the next combobox population will append to the existing entries, and that looks odd.

The rest of the routine is exactly the same as the Form_Load subroutine except for lines 23 and 38. Line 23 sets the database pointer to the selected combobox entry from the screen, and line 38 makes that same entry appear in the drop-down list box.

```
1. Option Explicit
2. Dim MyRecordC As tSalesHistoryCountRecord
3. Dim CustRow As Integer

4. Private Sub MenuButton_Click()
5. Me.Hide
6. Unload Me
7. frmMenu.Show
8. End Sub

9. Private Sub CustAFComboBox1_Click()
10. CustRow = CustAFComboBox1.ListIndex
11. OpenSalesHistoryCountDatabase
12. CustAFComboBox1.Clear
13. PDBMoveFirst (dbSalesHistoryCount)
14. If PDBEOF(dbSalesHistoryCount) Then
15. CustAFComboBox1.AddItem "No Customers"
16. Else
17. While Not PDBEOF(dbSalesHistoryCount)
18. ReadSalesHistoryCountRecord MyRecordC
19. CustAFComboBox1.AddItem MyRecordC.CustomerName + " - " +
      MyRecordC.WineName
20. PDBMoveNext (dbSalesHistoryCount)
```

Continues

Listing 10.7 Sales History code.

```
Wend
End If
GoToSalesHistoryCountRecord CustRow
ReadSalesHistoryCountRecord MyRecordC
AFTextBox1.Text = MyRecordC.Jan
AFTextBox2.Text = MyRecordC.Feb
AFTextBox3.Text = MyRecordC.Mar
AFTextBox4.Text = MyRecordC.Apr
AFTextBox5.Text = MyRecordC.May
AFTextBox6.Text = MyRecordC.Jun
AFTextBox7.Text = MyRecordC.Jul
AFTextBox8.Text = MyRecordC.Aug
AFTextBox9.Text = MyRecordC.Sep
AFTextBox10.Text = MyRecordC.Oct
AFTextBox11.Text = MyRecordC.Nov
AFTextBox12.Text = MyRecordC.Dec
AFTextBox13.Text = MyRecordC.WineName
CustAFComboBox1.ListIndex = CustRow
CloseSalesHistoryCountDatabase
End Sub

Private Sub Form_Load()
OpenSalesHistoryCountDatabase
CustAFComboBox1.Clear
PDBMoveFirst (dbSalesHistoryCount)
If PDBEOF(dbSalesHistoryCount) Then
CustAFComboBox1.AddItem "No Customers"
Else
While Not PDBEOF(dbSalesHistoryCount)
ReadSalesHistoryCountRecord MyRecordC
CustAFComboBox1.AddItem MyRecordC.CustomerName + " - " +
    MyRecordC.WineName
PDBMoveNext (dbSalesHistoryCount)
Wend
End If
CustAFComboBox1.ListIndex = 0
CustRow = 0
GoToSalesHistoryCountRecord CustAFComboBox1.ListIndex
ReadSalesHistoryCountRecord MyRecordC
AFTextBox1.Text = MyRecordC.Jan
AFTextBox2.Text = MyRecordC.Feb
AFTextBox3.Text = MyRecordC.Mar
AFTextBox4.Text = MyRecordC.Apr
AFTextBox5.Text = MyRecordC.May
AFTextBox6.Text = MyRecordC.Jun
AFTextBox7.Text = MyRecordC.Jul
AFTextBox8.Text = MyRecordC.Aug
AFTextBox9.Text = MyRecordC.Sep
AFTextBox10.Text = MyRecordC.Oct
AFTextBox11.Text = MyRecordC.Nov
AFTextBox12.Text = MyRecordC.Dec
AFTextBox13.Text = MyRecordC.WineName
CloseSalesHistoryCountDatabase
End Sub
```

Listing 10.7 Sales History code. *(Continued)*

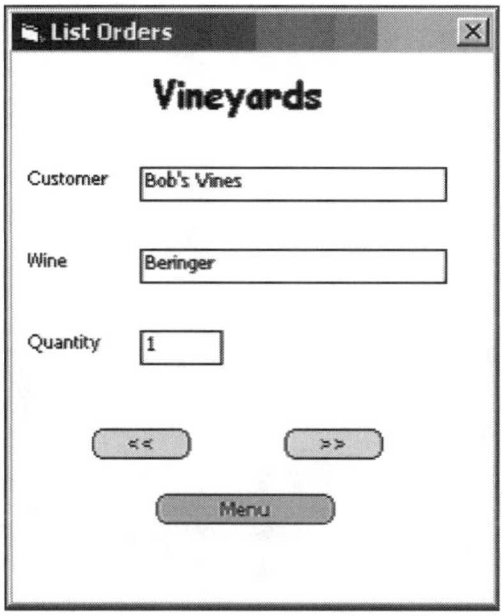

Figure 10.26 List orders menu.

This next main menu option is the viewing of all new orders placed since the last desktop synchronization. Referring back to the AppForge Conduit section, the Orders table was set to perform two-way synchronization. This means that when synchronization is initiated between the wireless device and the desktop, all entries on the device are downloaded to the desktop, and all new entries on the desktop are placed on the device.

This screen (see Figure 10.26) is nearly identical to the Wines menu program; it has scrolling capabilities and is nearly line for line the same code.

If you compare this program (Listing 10.8) to the Wines program, they are basically the same except for the name changes for the database, database fields, and screen variables fields.

This next screen has the largest program at 135 lines (Listing 10.9). The idea behind this order-placing program is simple. All customers are listed in the customer drop-down box, all wines are listed in their own drop-down box, and for each wine, the total number of available cases is shown. To place an order, select the desired customer, select the desired wine, and tap on the number of cases to order. The number of available cases will drive the radio button choices. As in Figure 10.27, if only one case is available, the screen will not allow the user to select more than one case. If three cases are available, the user can select one, two, or three cases, but no more. At this time, the number of cases and the number of orders placed are independent. Extra logic would have to be added to take the orders under consideration so as not to misrepresent the inventory levels.

```
Option Explicit
Dim MyRecord As tOrdersRecord

Private Sub PreviousButton_Click()
PDBMovePrev (dbOrders)
If Not PDBBOF(dbOrders) Then
DisplayData
End If
End Sub

Private Sub NextButton_Click()
PDBMoveNext (dbOrders)
If Not PDBEOF(dbOrders) Then
DisplayData
End If
End Sub

Private Sub MenuButton_Click()
CloseOrdersDatabase
Me.Hide
Unload Me
frmMenu.Show
End Sub

Private Sub Form_Load()
OpenOrdersDatabase
PDBSetSortFields dbOrders, tOrdersDatabaseFields
    .CustomerName_Field
PDBMoveFirst (dbOrders)
DisplayData
End Sub

Public Sub DisplayData()
If PDBNumRecords(dbOrders) > 0 Then
ReadOrdersRecord MyRecord
AFTextBox1.Text = MyRecord.CustomerName
AFTextBox2.Text = MyRecord.WineName
AFTextBox3.Text = MyRecord.Cases
Else
AFTextBox1.Text = "None"
AFTextBox2.Text = ""
AFTextBox3.Text = ""
End If
End Sub
```

Listing 10.8 List Orders code.

This program has all the previous logic incorporated into it as well as the ability to access multiple databases. Because each table is really a database, the program must open, read, and close the database for each table being accessed. Let's have a look.

Figure 10.27 Place an Order screen.

Lines 2, 3, and 4 define variables for Customers, Wines, and Orders. Lines 5 and 6 define variables the same as in the previous program.

Again, if the Menu button is clicked, application flow returns to the main menu screen as all other Menu button subroutines in our application.

Starting at the Form_Load again (lines 61 through 94), the first bit of processing is to open the customer database, clear the customer combobox, and set the cursor to the first customer record in the database. This then loops through the customers, loading up the customer combobox before closing the customer database and setting the viewable customer name to the first customer record. Lines 77 through 96 basically do the same thing for the Wines table, with the added exception of setting the number of total wine cases field from the wine inventory field in the database for the first wine in the database. A subroutine called SetButtons (lines 97 through 135) is called from line 95 and the radio buttons are set. If, for example, only one case of that particular wine is available, then all radio buttons other than the first are disabled, therefore preventing the user from selecting more cases than are available.

When a customer name is selected, the CustAFComboBox1_Click subroutine is invoked. This sets the CustRow variable to the customer name selected by the user in the dropdown box (line 13).

The user can also tap on a wine from the wine list, which executes the WineAFComboBox1_Click subroutine at lines 52 through 60. Within this subroutine, the wine database is opened, the specific wine is selected (based on its entry in the combobox), the index value is stored in the WineRow variable, and the database record is read. Then, the buttons are again set in accordance with the number of cases from the Inventory field in the database record. Finally, the Wine database is closed.

Now the two important variables, CustRow and WineRow are set, allowing the order to be placed. Lines 15 through 51 are for the OrderButton_Click subroutine. When the end user taps on the Place Order button, this subroutine is invoked. To start if no radio button has been selected, the screen is unloaded and completely reloaded from scratch. Since the customer and wine are set to default values, all that is required is to click on a radio button to select the number of cases of wine to order. Therefore, if a radio button is selected the process continues.

The Customers database is opened (line 33), the database pointer is set to whatever is contained in CustRow (line 34), and the appropriate record is read. The customer name is then placed in the MyRecordO.CustomerName variable field before the Customers database is closed on line 37. The same idea applies to the Wines database from line 38 through 42. MyRecordO.WineName field is set to the selected wine name on the screen.

Finally, with all the Orders fields filled in, the PDBCreatedRecordBySchema command is executed on line 43, and line 44 writes the new record to the database. Line 45 then updates the orders database, and line 46 closes the Orders database to commit the new entry.

That's it! A new order has been inserted into the database, and application flow returns to the Main Menu screen for ongoing processing.

```
1. Option Explicit
2. Dim MyRecordC As tCustomersRecord
3. Dim MyRecordW As tWinesRecord
4. Dim MyRecordO As tOrdersRecord
5. Dim CustRow As Integer
6. Dim WineRow As Integer

7. Private Sub MenuButton_Click()
8. frmOrder.Hide
9. Unload Me
10. frmMenu.Show
11. End Sub

12. Private Sub CustAFComboBox1_Click()
13. CustRow = CustAFComboBox1.ListIndex
14. End Sub
```

Listing 10.9 New Order code.

```
Private Sub OrderButton_Click()
MyRecordO.Cases = 0
If AFRadioButton1.Value = True Then
MyRecordO.Cases = 1
ElseIf AFRadioButton2.Value = True Then
MyRecordO.Cases = 2
ElseIf AFRadioButton3.Value = True Then
MyRecordO.Cases = 3
ElseIf AFRadioButton4.Value = True Then
MyRecordO.Cases = 4
ElseIf AFRadioButton5.Value = True Then
MyRecordO.Cases = 5
Else
Unload Me
frmOrder.Show
End If

If MyRecordO.Cases <> 0 Then
OpenOrdersDatabase

OpenCustomersDatabase
GoToCustomersRecord CustRow
ReadCustomersRecord MyRecordC
MyRecordO.CustomerName = MyRecordC.CustomerName
CloseCustomersDatabase

OpenWinesDatabase
GoToWinesRecord WineRow
ReadWinesRecord MyRecordW
MyRecordO.WineName = MyRecordW.WineName
CloseWinesDatabase

PDBCreateRecordBySchema dbOrders
WriteOrdersRecord MyRecordO
PDBUpdateRecord dbOrders
CloseOrdersDatabase

Me.Hide
Unload Me
frmMenu.Show
End If
End Sub

Private Sub WineAFComboBox1_Click()
OpenWinesDatabase
GoToWinesRecord WineAFComboBox1.ListIndex
WineRow = WineAFComboBox1.ListIndex
ReadWinesRecord MyRecordW
TotalCases.Text = MyRecordW.Inventory
SetButtons
CloseWinesDatabase
End Sub
```

Continues

Listing 10.9 *(Continued)*

```
Private Sub Form_Load()
OpenCustomersDatabase
CustAFComboBox1.Clear
PDBMoveFirst (dbCustomers)
If PDBEOF(dbCustomers) Then
CustAFComboBox1.AddItem "No Customers"
Else
While Not PDBEOF(dbCustomers)
ReadCustomersRecord MyRecordC
CustAFComboBox1.AddItem MyRecordC.CustomerName
PDBMoveNext (dbCustomers)
Wend
End If
CloseCustomersDatabase
CustAFComboBox1.ListIndex = 0
CustRow = 0

OpenWinesDatabase
WineAFComboBox1.Clear
PDBMoveFirst (dbWines)
If PDBEOF(dbWines) Then
WineAFComboBox1.AddItem "No Wines"
Else
While Not PDBEOF(dbWines)
ReadWinesRecord MyRecordW
WineAFComboBox1.AddItem MyRecordW.WineName
PDBMoveNext (dbWines)
Wend
End If
PDBMoveFirst (dbWines)
ReadWinesRecord MyRecordW
TotalCases.Text = MyRecordW.Inventory
CloseWinesDatabase
WineAFComboBox1.ListIndex = 0
WineRow = 0
SetButtons
End Sub

Public Sub SetButtons()
If TotalCases.Text = "0" Then
AFRadioButton1.Enabled = False
AFRadioButton2.Enabled = False
AFRadioButton3.Enabled = False
AFRadioButton4.Enabled = False
AFRadioButton5.Enabled = False
ElseIf TotalCases.Text = "1" Then
AFRadioButton1.Enabled = True
AFRadioButton2.Enabled = False
AFRadioButton3.Enabled = False
AFRadioButton4.Enabled = False
AFRadioButton5.Enabled = False
ElseIf TotalCases.Text = "2" Then
AFRadioButton1.Enabled = True
```

Listing 10.9 New Order code. *(Continued)*

```
AFRadioButton2.Enabled = True
AFRadioButton3.Enabled = False
AFRadioButton4.Enabled = False
AFRadioButton5.Enabled = False
ElseIf TotalCases.Text = "3" Then
AFRadioButton1.Enabled = True
AFRadioButton2.Enabled = True
AFRadioButton3.Enabled = True
AFRadioButton4.Enabled = False
AFRadioButton5.Enabled = False
ElseIf TotalCases.Text = "4" Then
AFRadioButton1.Enabled = True
AFRadioButton2.Enabled = True
AFRadioButton3.Enabled = True
AFRadioButton4.Enabled = True
AFRadioButton5.Enabled = False
Else
AFRadioButton1.Enabled = True
AFRadioButton2.Enabled = True
AFRadioButton3.Enabled = True
AFRadioButton4.Enabled = True
AFRadioButton5.Enabled = True
End If
End Sub
```

Listing 10.9 New Order code.

The last option on the Main Menu screen, other than the Exit button, is the *About* option. Once the About button is tapped, flow moves to the frmAbout screen, which simply displays a static screen with a hardcoded text area (see Figure 10.28). The screen has only one option, which is to return to the main menu via the Menu button.

The underlying code is very simple as shown in Listing 10.10.

That's the entire Vineyards application in a nutshell—hope it made sense and you can adapt this example for your further development.

```
Option Explicit
Private Sub MenuButton_Click()
Me.Hide
Unload Me
frmMenu.Show
End Sub
```

Listing 10.10 About screen code.

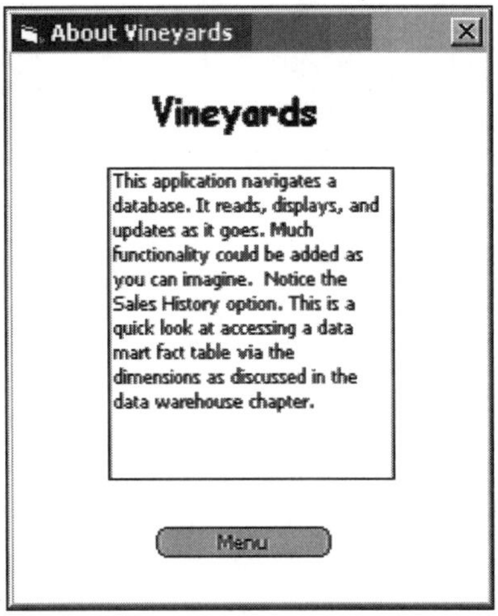

Figure 10.28 The About screen.

Final Thoughts

Sybase was one of the first database vendors to supply a full enterprise extension onto PDA units. They are fully dedicated to the advancement of wireless mobile development and are constantly coming up with strategies to help the marketplace move forward. With Sybase PDA database and synchronization technology, corporations can extend their enterprise information to the mobile workforce.

CHAPTER 11

IBM

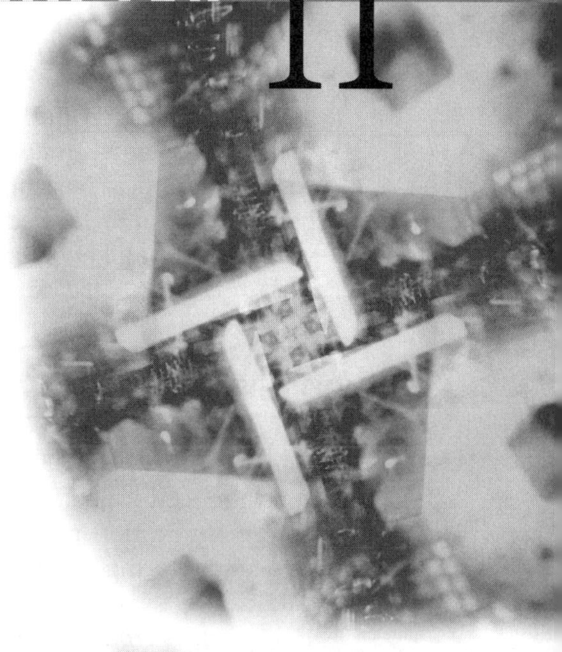

Throughout this chapter we will introduce IBM's offering in the portable database arena, DB2 Everyplace version 7.2.1. It is not a surprise that the name starts with the name of the "older brother" DB2, but the extension ("Everyplace") in the name tells us all about the nature of this product. As you will see in this chapter, IBM did a great job in the number of platforms supported. Similar to the other products in this category, DB2 Everyplace supports the subset of the SQL 99 standard. However, before we start let's see what components are required to build the examples provided in this chapter. Each of the components is available as a free evaluation download from the IBM Web site (after registration):

- IBM DB2 Everyplace Enterprise Edition version 7.2.1 (available for download from www.ibm.com/software/data/db2/everyplace/downloads.html)
- IBM DB2 Everyplace Mobile Application Builder version 7.2.1 (available for download from www.ibm.com/software/data/db2/everyplace/downloads.html)
- Microsoft Embedded Visual Tools version 3.0 (available for download at www.microsoft.com/mobile/downloads/emvt30.asp)
- Microsoft Pocket PC 2002 SDK (available for download at www.microsoft.com/mobile/developer/downloads/ppcsdk2002.asp)
- Active Sync 3.5 (minimum version is 3.1—delivered with Compaq iPAQ Pocket PC)

As you will see, we will need two different sets of application development tools to develop cross-platform applications. You will use IBM DB2 Everyplace Mobile Application Builder for Palm OS applications, because you can produce database-driven applications without ever writing a single line of code. To create applications for Windows CE-based devices, you will need eVB, a part of Microsoft's Embedded Visual Tools. IBM has provided us (developers) with a set of Call Level Interface (CLI) APIs that can be used to connect, query, and update IBM DB2 databases on portable devices.

Overview

One of the relational databases that support most platforms and operating systems has been scaled down for use on portable devices. IBM's DB2 Everyplace Enterprise Edition provides a local data store on the mobile or embedded device for storing relational data from anywhere in the enterprise. Relational data can be synchronized to the device from other data sources (see Figure 11.1), such as DB2 (all platforms supported Unix, OS/2, Windows NT, DB2 for OS/390, DB2 for AS/400), as well to the data sources provided by Oracle, Microsoft SQL Server, and Sybase.

Support from the palm-sized device to the mainframe while maintaining core product (SQL 99 standard) compatibility is a great achievement. However, it does not stop there. Support provided for different PDA platforms is impressive:

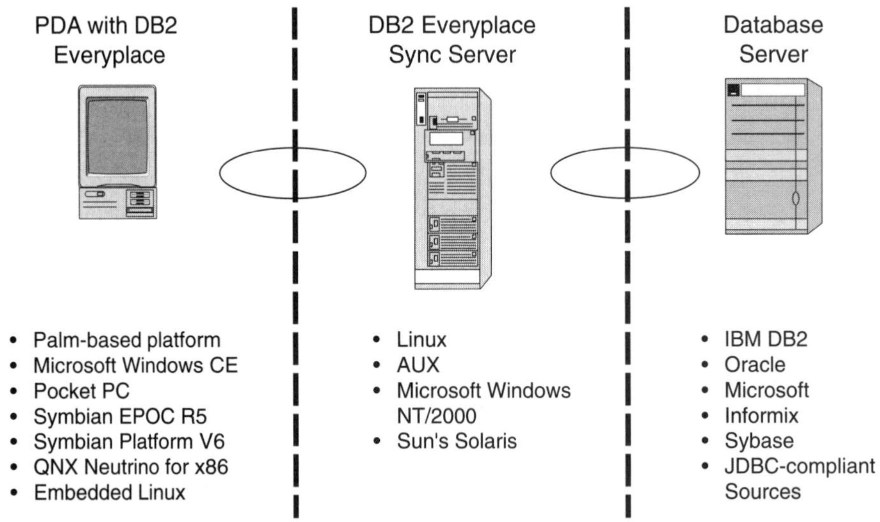

Figure 11.1 DB2 servers and platforms.

- Palm OS
- Symbian OS version 6
- EPOC Release 5
- Windows CE/Pocket PC
- QNX Neutrino
- Linux
- Embedded Linux devices

The supported features and technical characteristics in the 137KB of memory footprint are just as impressive:

- Supports single-byte character sets, double character sets, and UNICODE.
- Data access through JDBC and DB2 CLI/ODBC C/C++ API.
- SQL functionality including transaction processing, multi-table joins.
- INSERT with sub-select, IN list, RTRIM, and LENGTH.
- Advanced indexing, including bi-directional index scanning for efficient updates and queries using decreased storage.
- Data Definition Language (DDL) supported commands are CREATE TABLE, DROP TABLE, CREATE INDEX and DROP INDEX.
- Data Manipulation Language (DML) supported commands are INSERT, DELETE, UPDATE, SELECT, Scrollable Cursor, JOIN, GROUP BY, ORDER BY, ASC, and DESC.
- Expression and aggregate functions are also supported.
- CHECK constraints, DEFAULT VALUE, multiple-column PRIMARY KEY, FOREIGN KEY.
- Supported data types are small integer, integer, decimal, char, varchar, Binary Large Object (BLOB), date, time, and timestamp.
- Command-line processor supporting data IMPORT/EXPORT.
- Secondary storage is supported for the following PDA components:
    - Compact Flash (CF)
    - Sony Memory Stick
    - Secure Digital (SD) Card,
    - Multimedia Card (MMC)
    - IBM Micro drive
- Columns per table: 128
- Indexes per table: 15
- Bi-directional, multi-attribute, multipurpose indexes.

- Number of tables limited only by available memory.
- Maximum length of SQL statement: 2KB.

The combination of covered platforms, operating systems, and built-in features makes DB2 an excellent candidate for large enterprises where a mixture of different platform and OS environments usually exists.

Major components of IBM DB2 Everyplace are:

- IBM DB2 Everyplace database on the portable device
- IBM DB2 Everyplace Sync Server
- IBM DB2 Everyplace Sync Client

System Tables

A database engine creates and maintains system catalog and base tables. DB2 Everyplace system tables are "in charge" of keeping the records of tables, columns, and primary-foreign key relationships. From the development perspective, it is important to know that occasionally system tables should be queried. It is also important to understand that standard SQL statements cannot modify system tables. Table 11.1 is a list of the system tables with a short overview of the table structures.

As you can see from the name (DB2eSYSTABLES), this table will contain one row for each table created in the system.

Table 11.1 DB2eSYSTABLES

NAME	DATA TYPE	LENGTH	DESCRIPTION
TNAME	VARCHAR	19	Name of the table
NUMCOLS	INTEGER	4	Number of columns
FLAGS	INTEGER	4	Internal use only
NUMKEY	INTEGER	4	Number of columns in the primary key
CHK	BLOB	512	Internal use only
IDXINFO	BLOB	700	Internal use only
NUMREFS	INTEGER	4	Primary and foreign keys
F_ID	INTEGER	4	Internal use only
PD	BLOB	256	Internal use only

Table 11.2 DB2eSYSCOLUMNS

NAME	DATA TYPE	LENGTH	DESCRIPTION
CNAME	VARCHAR	19	Name of the column
TNAME	VARCHAR	19	Name of the table
TYPE	INTEGER	4	Data type
ATTR	INTEGER	4	Field attributes—Internal use only
LENGTH	INTEGER	4	Length of column
POS	INTEGER	4	Column number
FLAGS	INTEGER	4	Internal use only
KEYSEQ	INTEGER	4	Ordinal position of the column in the primary key
SCALE	INTEGER	4	Decimal column scale
DEF	VARCHAR	128	Default value

DB2eSYSCOLUMNS table (Table 11.2) contains one row for each column that exists in the tables defined in the system.

DB2eSYSRELS (Table 11.3) contains one row for each referential constraint that is defined.

As you will see from the example provided in this chapter, the database manager creates system tables automatically once we create initial user tables. Any additional user tables will use already existing system tables.

Table 11.3 DB2eSYSRELS

NAME	DATA TYPE	LENGTH	DESCRIPTION
CNAME	VARCHAR	19	Name of the column
RMD_ID	INTEGER	4	Primary and foreign key
PKTABLE_NAME	VARCHAR	19	Parent table name
PKCOLUMN_NAME	VARCHAR	19	Primary key column in parent table
FKTABLE_NAME	VARCHAR	19	Child table name
FKCOLUMN_NAME	VARCHAR	19	Foreign key column in child table
ORDINAL_POSITION	INTEGER	4	Ordinal position of the column in the foreign key reference

Limitations

Some of the features of DB2 Everyplace were reserved for the "big" systems just couple of years ago. You could ask the question, "Who wants to store 4GB of data on a portable device? And even if I have that much information to store, how can I put gigabytes of data on a portable device?" IBM already makes "solid state" (storage devices based on storing the data into large amounts of non-erasable memory rather than on movable disk components) drives that can store gigabytes of data. For example, we could walk into our local library to download a book to our portable device instead of borrowing the physical book. Some books are rather large, and with that much space, you could borrow two or three books at a time.

Let's now look at some of the limitations that developers and database architects will encounter with the DB2 Everyplace database engine:

- Maximum table size (on a 32-bit system): 16 million pages at 512 bytes (4GB)
- Maximum number of tables in a data store: 65535
- Maximum number of indices on a table: 15
- Maximum number of foreign keys in a table: 8
- Maximum number of columns in an index: 8
- Maximum number of columns in a primary key: 8
- Maximum length of SQL statement: 64 kilobytes
- Maximum number of connections to a data store path: 1
- Maximum number of rows in a table is limited only by table size
- Maximum length for a CHAR column: 32 kilobytes
- Maximum length for VARCHAR or BLOB column: 32 kilobytes
- Maximum cumulative length for row's fixed-length columns: 64 kilobytes
- Maximum number of statement handles per connection: 20
- Maximum length of default values when inserting: 128 bytes
- Maximum length of check constraints: 512 bytes
- Maximum size of decimals: 31 digits

With the new generations of portable devices just around the corner and new and improved versions of DB2 Everyplace, we should expect the aforementioned limits to be raised in the near future.

Installation on the Windows Workstation

To install DB2 Everyplace on the Windows workstation, you will need a system that satisfies the minimum hardware requirements:

- Intel Pentium II (or comparable AMD processor) or better
- Support for the communications (USB, serial port) adapter
- A minimum of 256MB of memory (512MB is recommended)
- At least 175MB of free disk space to install the DB2 Everyplace database
- Sync Server

Numerous parts of DB2 Everyplace are Java based (especially administrative tools). Unless you have a powerful CPU and the recommended amount of memory, you will notice slowdowns. This is true especially when all of the components are installed on the same workstation.

A large selection of supported operating systems is available. For the Windows-based systems, you can use either Windows NT or Windows 2000.

The list of software packages that you will need on the workstation is fairly impressive and a medium to advanced level of comprehensive understanding of DB2 inner workings and the DB2 environment will be required for the successful installation and running of the complete system.

Let's see what is required:

- IBM DB2 Universal Database version 7.2
- DB2 Everyplace Database version 7.2.1
- DB2 Everyplace Sync Server version 7.2.1
- Web server with Java Servlet API 2.0 support (A basic version of the IBM's WebSphere WebServer is included with DB2 Everyplace Enterprise Edition.)
- Workstation to mobile device connection software such as Palm HotSync or CE ActiveSync.

The next step will be to install DB2 Everyplace on to your workstation. At this point, you should have the installation file downloaded to your PC from the IBM DB2 Everyplace site. The filename for the English version of the product should be DB2Everyplace721fp-ee-enu.exe. Locate this file on your file system and double-click on the filename. You will be asked to select the language to use for this installation. Since English is the default in this case, just click OK.

When DB2 Everyplace opens, select the Install button because you have to read hardware and operating system requirements. If you decide to click on one of the other buttons on this screen, you will be provided with a text document, which explains compatibility and system requirements, as well as the latest application notes.

The following three screens allow you to select components that you would like to install, the location of the DB2 Everyplace installation, and the folder on your Start menu that will be used for the home DB2 Everyplace shortcuts.

You can click Next on each of the screens because the default options will be appropriate choices. On the next screen, you will be presented with a choice of mobile platforms to which DB2 Everyplace can be deployed. In our case, select Palm, Windows CE, and Win32 platforms.

When you have checked the appropriate boxes, click Next. At this point, you have almost completed installation of the application. If you want to install Sync Server sample application, you will have to click the Finish button. It is important to note that the Sync Server and sample application for the synchronization and replication is only available in the Enterprise edition of DB2 Everyplace, so make sure that you have downloaded the correct version from IBM's site. There are three DB2 Everyplace Sync Server samples: VNURSE, M_VNURSE, and M_VN2. These cover all of the functionality aspects that DB2 Everyplace offers: relational database engine, synchronization across the Internet or local LAN, and portability. If you elect not to install these samples and the appropriate replication and synchronization definitions now, you can do so later. However, it is much easier to perform this task through the graphical user interface (GUI) provided by the DB2 Everyplace installation application than with the batch files.

Installation on Mobile Device

When you have restarted your workstation (this step is not required, but it is recommended), you have the straightforward task of propagating (installing) required components to the mobile device of your choice. Well, at least that's the choice that we selected during the installation.

Select Start->DB2 Everyplace->Install on the mobile device option from your Start menu. Select the appropriate mobile device operating system (in our case, that will be "Windows CE devices") and proceed to the next screen. Here you will have to make a choice on what components (by default all of them are selected) you would like to install. Leave the default values checked and proceed by clicking Apply. You will go through a couple of device-related messages, which will instruct you to connect your device and log on, and will notify you when that task has been completed.

Informational screens will be shown to you for the duration of the install process. When this process is completed, you will be returned to the screen where you had to select the desired DB2 Everyplace components.

Congratulations! At this point you have DB2 Everyplace installed on your mobile device. Do not forget that in order to use replication and synchronization, you will also need DB2 Universal Database (UDB) installed on your system.

Development Tools

DB2 offers support for a variety of development environments. Selection of the tools you can use is driven by the fact that DB2 supports so many different platforms. Many commercially available tools, such as CodeWarrior, DB2 Everyplace Mobile Application Builder, Visual Age Micro Edition, or GNU Software Developers Kit will produce excellent results. These tools are based on C/C++ and Java, but what about Visual Basic based tools? Well, Embedded Visual Basic (eVB) is supported via a CLI/ODBC API interface provided by IBM. In this chapter, we will create an application based on CLI/ODBC interface using eVB, but in Chapter 7, "Mobile Application Development Tools," we also discussed DB2 Everyplace Mobile Application Builder because this development tool can produce results on Palm-based devices with minimum or no code at all.

Airplane Tester Application

Our sample application, which will be presented in this chapter, is actually combined from two separate applications. One will create the required tables (and appropriate system catalogs and tables) and populate them with the sample data, and the second will execute a simple SQL statement against a single DB2 Everyplace table. It will also execute a more complex SQL statement in which we will join results from two tables and explain the principle of drilling down the data. All of the code will be created based on Microsoft's eVB, and interfacing with DB2 Everyplace will be performed through CLI/ODBC. If you would like to see a complete list of all available CLI/ODBC functions, please review Appendix C, "DB2 CLI/ODBC Functions."

It is important to note that IBM did everything possible to cover all the different development tastes and desires. If you are proficient in one of the major

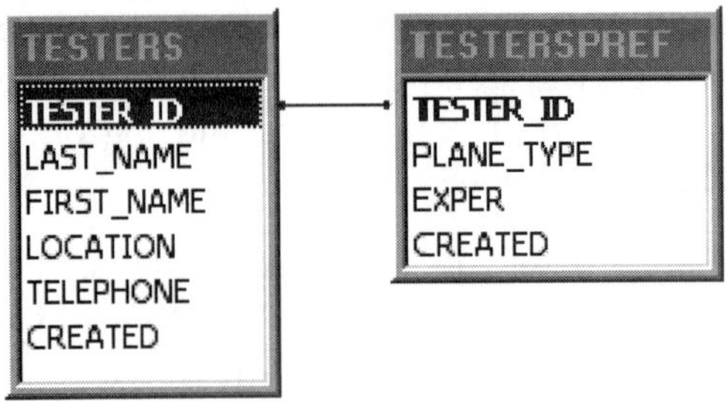

Figure 11.2 Tables required for sample application.

development environments, you will be able to create great PDA applications. Even more, they are one of the rare vendors that will enable you to create applications for the two major groups of PDAs (Palm and CE-based PDAs) with the tools that they provide. Our example will completely reside on the portable device (application and database), but at the end of this chapter we will explain how replication and synchronization can be achieved. In order to proceed with the sample applications, we will assume that you have installed the sample DB2 Everyplace applications (explained previously in this chapter) on your portable device. This is important because installation of the sample applications installs the DB2 Everyplace database components onto your portable device (in our case, that will be a Windows CE-based device).

The idea behind the sample application is quite simple: Provide a portable database and application with a listing of test pilots specialized in commercial and military planes. We will have only two tables (see Figure 11.2) in our design: A Testers table that will hold a majority of the personal information, and a TestersPref table that will hold information about experience and the type of airplane in which the tester is an expert.

Create Testers Tables

Before we can even start thinking about listing our pilot experts on the screen of your portable device, we must create a DB2 Everyplace database. This application is represented with the simple flowchart in Figure 11.3. Logically, there are only two components to this part of the application: Create tables successfully, or table creation failed.

The starting point for this application is shown in the code in Listing 11.1.

The form that the end user will see is shown in Figure 11.4. It is clear that further action will be taken by pressing the Create Tables button. That simple action starts the whole chain of DB2 Everyplace CLI/ODBC API statements that will result in the creation of the desired tables and system catalogs.

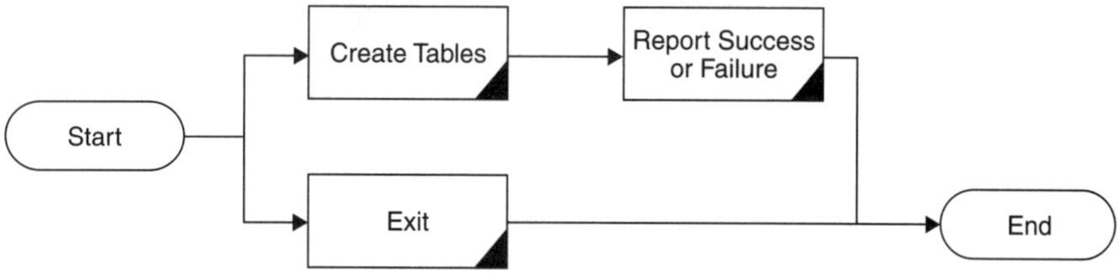

Figure 11.3 Application flow for database and table creation.

```
'-----------------------------------------------------------------
'      Main Form for Tester demo application
'
'           Form Name : frmCreateTesters
'           Creator : Developer
'           TypeID    : PRG
'
'         Created Time : 4/5/2002 6:19:42 AM
''----------------------------------------------------------------
Private Sub Command1_Click()
    ' Start createin the tables
    rc = MakeTesters()
End Sub
Private Sub Command2_Click()
    ' Close the form and terminate application
    App.End
    Stop
End Sub
Private Sub Form_OKClick()
    App.End
End Sub
```

Listing 11.1 Main form for database and table creation.

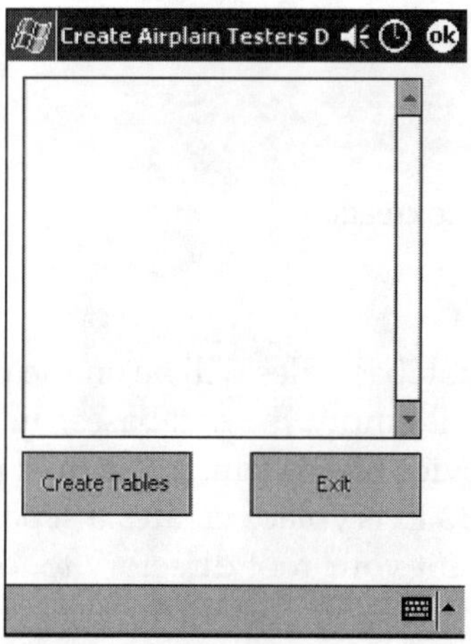

Figure 11.4 Database and table screenshot.

In order to understand the flow of CLI/ODBC interface, let's review the required stages in the table creation process, step-by-step. All of the required code is contained in one Basic module called modFunctions.

```
1. `----------------------------------------------------------------
2. `     Basic functions module for Testers application
3. `
4. `          Module Name : modFunctions
5. `          Creator : Developer
6. `          TypeID     : PRG
7. `
8. `       Created Time : 4/5/2002 6:19:42 AM
9. `'---------------------------------------------------------------
10. Public henv    As Long
11. Public hdbc    As Long
12. Public hstmt   As Long
13. Public rc      As Integer
14. Public dbloc As String
15. Public userid As String
16. Public pass   As String
17. Public Function MakeTesters() As Integer
18. Dim i As Integer
19. Dim data As String
20. Dim ResultRows As Long
21. On Error Resume Next
```

Listing 11.2 Main module for database and table creation.

Lines of code (1 through 13 in Listing 11.2) are in place that will assign public variables that will be used throughout the application.

```
22. dbloc = "\\windows\programs\database\"
23. ` Userid and password do not exists in the CE version
24. userid = "nothing"
25. pass = "nothing"
```

Listing 11.3 Main module for database and table creation.

Line 22 specifies where the location of the database files will be on the portable (Windows CE) device. If that variable dbloc is empty, the databases, by default, will be located in the root of the portable device file system. It is a good practice to specify this value (do not forget that DB2 Everyplace creates a separate file for each new table), because you do not want your root directory to look congested if you have 10 or more tables. It is also important to note that DB2 Everyplace does not support user IDs and passwords. That is the reason why, on lines 24 and 25 in Listing 11.3, we can assign a random value. In this case, we will assign "nothing" values to both variables.

```
26. crtStmt1 = "CREATE TABLE TESTERS (TESTER_ID INT, LAST_NAME
      CHAR(30), FIRST_NAME CHAR(30), LOCATION CHAR(30), TELEPHONE
      CHAR(30), CREATED TIMESTAMP)"
27. crtStmt2 = "CREATE TABLE TESTERSPREF (TESTER_ID INT, PLANE_TYPE
      CHAR(30), EXPER CHAR(2),CREATED TIMESTAMP)"
28. insStmt1 = "INSERT INTO TESTERS VALUES(1, 'VUJOSEVIC', 'SRDJAN',
      'Toronto', '416-243-7381', CURRENT TIMESTAMP)"
29. insStmt2 = "INSERT INTO TESTERS VALUES(2, 'LABERGE', 'ROBERT',
      'Montreal', '999-888-7777', CURRENT TIMESTAMP)"
30. insStmt3 = "INSERT INTO TESTERS VALUES(3, 'MIG', 'ROOSTER',
      'Moscow', '011-1234-345-3456', CURRENT TIMESTAMP)"
31. insStmt4 = "INSERT INTO TESTERSPREF VALUES(1, 'Boeing 747', '10',
      CURRENT TIMESTAMP)"
32. insStmt5 = "INSERT INTO TESTERSPREF VALUES(2, 'DC10', '6', CURRENT
      TIMESTAMP)"
33. insStmt6 = "INSERT INTO TESTERSPREF VALUES(3, 'MIG21', '25',
      CURRENT TIMESTAMP)"
typedef struct {
LocalID nextRecordListID;
UInt16 numRecords;
UInt16 firstEntry;
} RecordListType;
```

Listing 11.4 Main module for database and table creation.

We are at that point in the code (lines 26 through 33 in Listing 11.4) where data definition language (DDL) statements (lines 26 and 27) and data manipulation statements (DML) statements (lines 28 through 33) are defined and stored in variables that will be executed later in this chapter.

Parts of the code that follow will serve to build, one step at the time, a connection to the database. In the end it will execute the SQL statements prepared on lines 26 though 33. In general terms, the process of executing the SQL statements in the CLI/ODBC environment is based on the following steps:

1. Allocate environment handle (lines 34 through 42)
2. Allocate database handle (lines 43 through 50)
3. Connect to database (lines 51 through 59)
4. Allocate SQL statement handle (lines 60 through 67)
5. Execute the statements (lines 68 through 131)
6. Disconnect from database (line 142)
7. Deallocate database handle (line 143)
8. Deallocate environment handle (line 144)

```
data = String(40, " ")
rc = SQLAllocHandle(SQL_HANDLE_ENV, SQL_NULL_HENV, henv)
If (rc <> 0) Then
frmCreateTesters.txtTesters.Text = frmCreateTesters.txtTesters
    .Text & " Error Allocating Env. Handle " & vbCrLf
rc = DB2Stop()
Exit Function
Else
frmCreateTesters.txtTesters.Text = frmCreateTesters.txtTesters
    .Text & " Allocated Environment Handle " & vbCrLf
End If

rc = SQLAllocHandle(SQL_HANDLE_DBC, henv, hdbc)
If (rc <> 0) Then
frmCreateTesters.txtTesters.Text = frmCreateTesters.txtTesters
    .Text & " Error Allocating Db. Handle " & vbCrLf
rc = DB2Stop()
Exit Function
Else
frmCreateTesters.txtTesters.Text = frmCreateTesters.txtTesters
    .Text & " Allocated Database Handle " & vbCrLf
End If
```

Listing 11.5 Main module for database and table creation.

```
' Connect to the database
rc = SQLConnect(hdbc, dbloc, SQL_NTS, userid, SQL_NTS, pass,
    SQL_NTS)
If (rc <> 0) Then
frmCreateTesters.txtTesters.Text = frmCreateTesters.txtTesters
    .Text & " Error Connecting to Db " & vbCrLf
rc = DB2Stop()
Exit Function
Else
frmCreateTesters.txtTesters.Text = frmCreateTesters.txtTesters
    .Text & " Connected to Database " & vbCrLf
End If

rc = SQLAllocHandle(SQL_HANDLE_STMT, hdbc, hstmt)
If (rc <> 0) Then
frmCreateTesters.txtTesters.Text = frmCreateTesters.txtTesters
    .Text & " Error Allocating Stmt. Handle " & vbCrLf
rc = DB2Stop()
Exit Function
Else
frmCreateTesters.txtTesters.Text = frmCreateTesters.txtTesters
    .Text & " Allocated Statement Handle " & vbCrLf
End If
```

Listing 11.6 Main module for database and table creation.

It seems complicated, but do not forget that the CLI/ODBC API interface is a low-level connectivity method to the portable DB2 Everyplace database. As such, it requires more steps to connect to the database and execute SQL statements than, for example, ADO connections in the desktop world.

Listings of available CLI/ODBC APIs can be found in Appendix C. Please note that each of the steps has error detection. If the "rc" (return value from the API function call) value is zero, it will indicate that everything is okay and that we can proceed to the next statement. If the returned value for "rc" is anything other than 0, we will have to close the connection (for example, lines of code 68 through 75).

```
68. rc = SQLExecDirect(hstmt, crtStmt1, SQL_NTS)
69. If (rc <> 0) Then
70. frmCreateTesters.txtTesters.Text = frmCreateTesters.txtTesters
        .Text & " Error Creating Table " & vbCrLf
71. rc = DB2Stop()
72. Exit Function
73. Else
74. frmCreateTesters.txtTesters.Text = frmCreateTesters.txtTesters
        .Text & " Table TESTERS Created " & vbCrLf
75. End If

76. rc = SQLExecDirect(hstmt, insStmt1, SQL_NTS)
77. If (rc <> 0) Then
78. frmCreateTesters.txtTesters.Text = frmCreateTesters.txtTesters
        .Text & " Error Inserting Rows " & vbCrLf
79. rc = DB2Stop()
80. Exit Function
81. Else
82. frmCreateTesters.txtTesters.Text = frmCreateTesters.txtTesters
        .Text & " Rows Inserted " & vbCrLf
83. End If

84. rc = SQLExecDirect(hstmt, insStmt2, SQL_NTS)
85. If (rc <> 0) Then
86. frmCreateTesters.txtTesters.Text = frmCreateTesters.txtTesters
        .Text & " Error Inserting Rows " & vbCrLf
87. rc = DB2Stop()
88. Exit Function
89. Else
90. frmCreateTesters.txtTesters.Text = frmCreateTesters.txtTesters
        .Text & " Rows Inserted " & vbCrLf
91. End If

92. rc = SQLExecDirect(hstmt, insStmt3, SQL_NTS)
93. If (rc <> 0) Then
94. frmCreateTesters.txtTesters.Text = frmCreateTesters.txtTesters
        .Text & " Error Inserting Rows " & vbCrLf
95. rc = DB2Stop()
```

Continues

Listing 11.7 Main module for database and table creation.

```
Exit Function
Else
frmCreateTesters.txtTesters.Text = frmCreateTesters.txtTesters
    .Text & " Rows Inserted " & vbCrLf
End If

rc = SQLExecDirect(hstmt, crtStmt2, SQL_NTS)
If (rc <> 0) Then
frmCreateTesters.txtTesters.Text = frmCreateTesters.txtTesters
    .Text & " Error Creating Table " & vbCrLf
rc = DB2Stop()
Exit Function
Else
frmCreateTesters.txtTesters.Text = frmCreateTesters.txtTesters
    .Text & " Table TESTERSPREF Created " & vbCrLf
End If

rc = SQLExecDirect(hstmt, insStmt4, SQL_NTS)
If (rc <> 0) Then
frmCreateTesters.txtTesters.Text = frmCreateTesters.txtTesters
    .Text & " Error Inserting Rows " & vbCrLf
rc = DB2Stop()
Exit Function
Else
frmCreateTesters.txtTesters.Text = frmCreateTesters.txtTesters
    .Text & " Rows Inserted " & vbCrLf
End If

rc = SQLExecDirect(hstmt, insStmt5, SQL_NTS)
If (rc <> 0) Then
frmCreateTesters.txtTesters.Text = frmCreateTesters.txtTesters
    .Text & " Error Inserting Rows " & vbCrLf
rc = DB2Stop()
Exit Function
Else
frmCreateTesters.txtTesters.Text = frmCreateTesters.txtTesters
    .Text & " Rows Inserted " & vbCrLf
End If

rc = SQLExecDirect(hstmt, insStmt6, SQL_NTS)
If (rc <> 0) Then
frmCreateTesters.txtTesters.Text = frmCreateTesters.txtTesters
    .Text & " Error Inserting Rows " & vbCrLf
rc = DB2Stop()
Exit Function
Else
frmCreateTesters.txtTesters.Text = frmCreateTesters.txtTesters
    .Text & " Rows Inserted " & vbCrLf
End If

rc = DB2Stop()
DB2eTest = 0
End Function
```

Listing 11.7 Main module for database and table creation. *(Continued)*

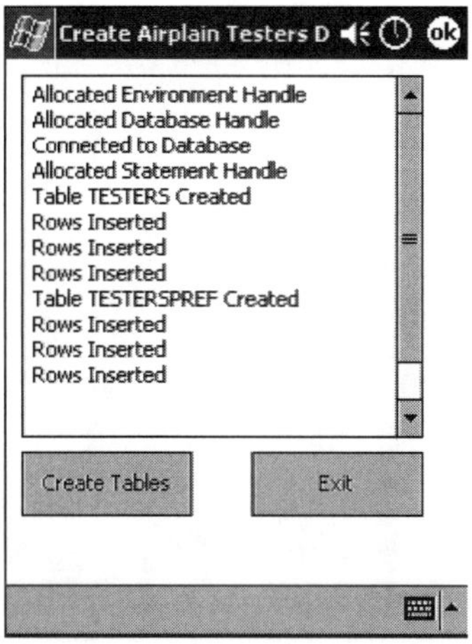

Figure 11.5 Database and table creation progress.

Code on lines 68 through 131 will finally execute the SQL DML and DDL statements that we prepared at the beginning of the code. If everything is okay, you should see the screen shown in Figure 11.5.

Since the process has been completed, we can invoke the function (DB2Stop) on lines 135 through 146 in Listing 11.8. This will close the database connection and deallocate database and environment handles.

```
135. Public Function DB2Stop() As Integer
136. On Error Resume Next
137. If rc = 0 Then
138. frmCreateTesters.txtTesters.Text = frmCreateTesters.txtTesters
        .Text & vbCrLf & "  Testers Sample tables created" & vbCrLf
139. Else
140. frmCreateTesters.txtTesters.Text = frmCreateTesters.txtTesters.
       Text & vbCrLf & "  Error(s) during table creation" & vbCrLf
141. End If
142. rc = SQLDisconnect(hdbc)
143. rc = SQLFreeHandle(SQL_HANDLE_DBC, hdbc)
144. rc = SQLFreeHandle(SQL_HANDLE_ENV, henv)
145. DB2Stop = 0
146. End Function
```

Listing 11.8 Main form for database and table creation.

What will happen if we tap the Create Tables button again? Because the tables are already in place (and we are not dropping tables prior to the attempt to create

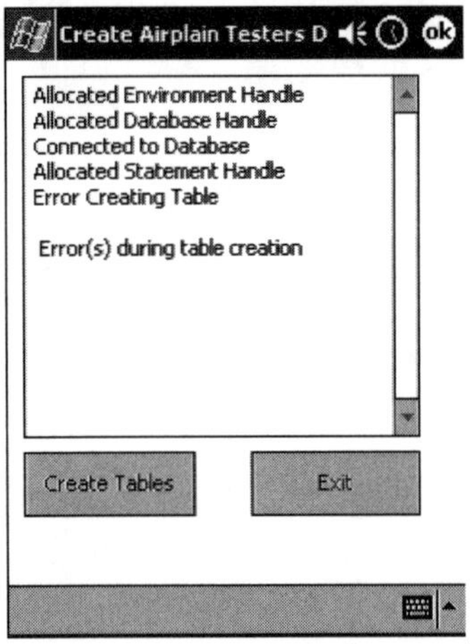

Figure 11.6 Error during creation.

them), our application will fail once it attempts to create tables. As we previously stated, the code on lines 68 through 75 in Listing 11.7 will not be able to create tables, and an "rc" value other than zero will force creation of an error message (in this case, "Error creating table"), and, in turn, it will force execution of the code that will close the connection to the database. You can see the result of the attempt to recreate the tables in Figure 11.6.

If you would like to see the list of created tables (systems and user) navigate to My Device->Windows->Programs->database (see Figure 11.7). It is interesting to note how different vendors have approached creation of the tables and concept of the database files. For example, Microsoft (even on the level of the portable device) creates a single database container file for all of the tables in

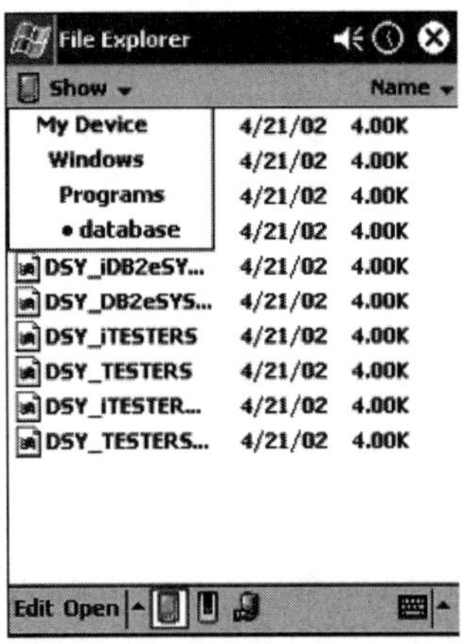

Figure 11.7 What files are created on Windows CE.

one database. IBM creates a separate file for each system and user table that will be required. In the IBM DB2 Everyplace case, it is immediately visible if tables are created successfully or not.

It is important to say that this part of our application also uses one more module (db2ecli.bas) provided by IBM, and it contains all of the required constant definitions as well as all of the required APIs and their appropriate declarations. This module is presented in Listing 11.11 and we will be using it in both applications.

Main Application

So, we have created required tables and populated them with sample data. Now, it is time to create an application that will query the data and present the results on meaningful screens for your end user. Let us review the application flow chart in Figure 11.8 for a moment.

We will provide the end user with the ability to perform simple and slightly more complex types of queries. The end user also will have the option to see the record details (when they are shown on the screen) by tapping the desired record in the record list.

The first screen we would like to present to the end user is shown in Figure 11.9. We have part of the screen allocated for the result set presentation in the

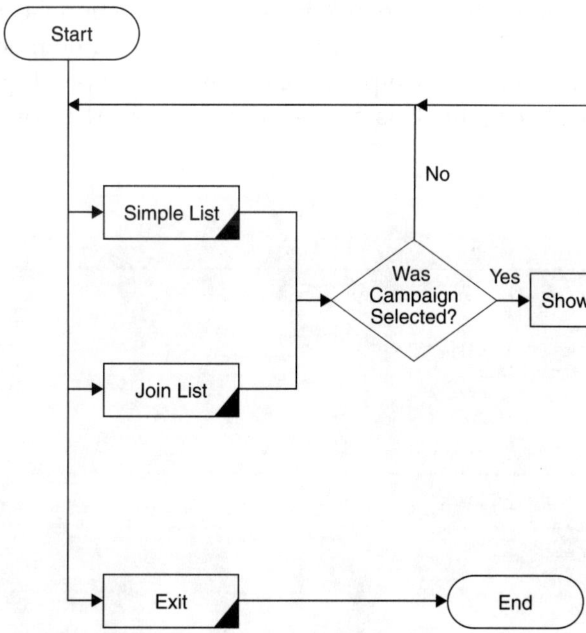

Figure 11.8 Testers sample application.

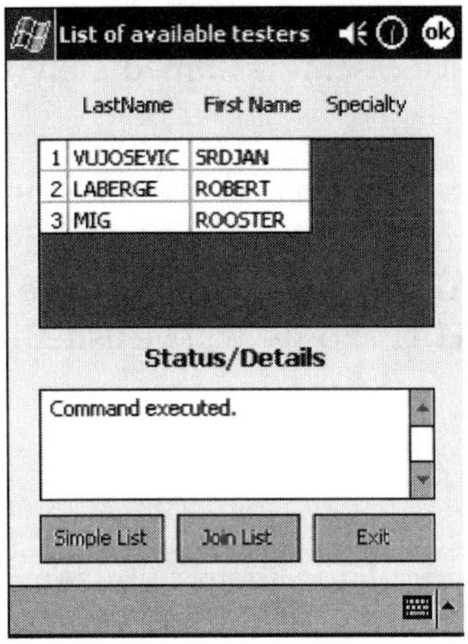

Figure 11.9 Simple query results.

form of a grid, an area of the screen where status messages will be presented (and at the same time, will be used for record details presentation), and a command button area.

The code in Listing 11.9 will be executed at the time of the application startup.

Click (or tap in the case of portable devices) on the Simple List button. The code on lines 31 through 36 in Listing 11.9 will be executed, global variable QueryType will be assigned the value of "simple", and function GetTesters() will be invoked. Function GetTesters is part of the Basic module modTesters and uses the global variable QueryType to distinguish the type of query we are going to use—simple or join. This can be seen on lines 31 through 35 in Listing 11.10.

```
1. '-------------------------------------------------------------
2. '     Main Form for Tester demo application
3. '
4. '          Form Name : frmTesters
5. '          Creator : Developer
6. '          TypeID    : PRG
7. '
8. '       Created Time : 4/5/2002 6:19:42 AM
9. ''-------------------------------------------------------------
10. Private Sub Command1_Click()
11. txtTesters.Text = ""
12. txtTesters.Refresh
13. QueryType = "join"
```

Listing 11.9 Main form for database query.

```
rc = GetTesters()
End Sub
Private Sub Exit_Click()
App.End
Stop
End Sub
Private Sub Form_Load()
DetailSel = ""
End Sub
Private Sub Form_OKClick()
App.End
End Sub
Private Sub Grid1_Click()
frmTesters.Grid1.Col = 0
DetailSel = frmTesters.Grid1.Text
rc = GetTesters()
End Sub
Private Sub Run_Click()
txtTesters.Text = ""
txtTesters.Refresh
QueryType = "simple"
rc = GetTesters()
End Sub
```

Listing 11.9 *(Continued)*

Opening the connection follows the same logic as in the previous example. When the SQL statement has been executed against the user tables, code on lines 73 through 91 in Listing 11.10 will populate the result grid on the screen of the portable device. If you clicked on the Simple List button, you will see the screen shown in Figure 11.9. If you selected the Join List screen, you should see the screen pictured in Figure 11.10. In a real-life production application, you

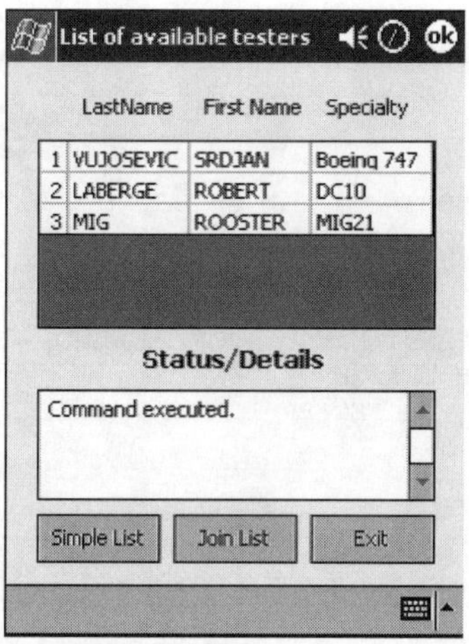

Figure 11.10 Result from multiple table query.

could select to hide the grid control until it is completely populated (much faster), but for this example, you will be able to see data populating cells in the grid control.

Note that the CLI/ODBC function used to retrieve data from the result set is SQLGetData() in combination with the SQLFetch() function (that will check to see if we reached the end of the result set).

```
'-------------------------------------------------------------
'     Basic functions module for Testers application
'
'          Module Name : modTesters
'          Creator : Developer
'          TypeID     : PRG
'
'       Created Time : 4/5/2002 6:19:42 AM
''------------------------------------------------------------
Public henv    As Long
Public hdbc    As Long
Public hstmt   As Long
Public rc      As Integer

Public dbloc As String
Public userid As String
Public pass   As String
Public QueryType As String
Public DetailSel As String

Public Function GetTesters() As Integer

Dim numCols As Integer
Dim i As Integer
Dim retLen As Long
Dim data As String
Dim selSQL As String
Dim ResultRows As Long

On Error Resume Next

dbloc = "\\windows\programs\database\"
userid = "nothing"
pass = "nothing"

selDetail = "SELECT a.TESTER_ID, a.LAST_NAME, a.FIRST_NAME,
    a.LOCATION, a.TELEPHONE, b.PLANE_TYPE FROM TESTERS a,
    TESTERSPREF b WHERE a.TESTER_ID = b.TESTER_ID AND a.TESTER_ID =
    " & CStr(DetailSel)

If QueryType = "simple" Then
selSQL = "SELECT TESTER_ID, LAST_NAME, FIRST_NAME FROM TESTERS"
```

Listing 11.10 Main module for database and table query.

```
Else
selSQL = "SELECT a.TESTER_ID, a.LAST_NAME, a.FIRST_NAME,
    b.PLANE_TYPE FROM TESTERS a, TESTERSPREF b WHERE a.TESTER_ID =
    b.TESTER_ID"
End If

data = String(40, " ")

rc = SQLAllocHandle(SQL_HANDLE_ENV, SQL_NULL_HENV, henv)
If (rc <> 0) Then
rc = DB2Stop()
Exit Function
End If

rc = SQLAllocHandle(SQL_HANDLE_DBC, henv, hdbc)
If (rc <> 0) Then
rc = DB2Stop()
Exit Function
End If

' Connect to the database
rc = SQLConnect(hdbc, dbloc, SQL_NTS, userid, SQL_NTS, pass,
    SQL_NTS)
If (rc <> 0) Then
rc = DB2Stop()
Exit Function
End If

rc = SQLAllocHandle(SQL_HANDLE_STMT, hdbc, hstmt)
If (rc <> 0) Then
rc = DB2Stop()
Exit Function
End If

If DetailSel = "" Then
rc = SQLExecDirect(hstmt, selSQL, SQL_NTS)
Else
rc = SQLExecDirect(hstmt, selDetail, SQL_NTS)
End If
If (rc <> 0) Then
rc = DB2Stop()
Exit Function
End If
rc = SQLNumResultCols(hstmt, numCols)
If (rc <> 0) Then
rc = DB2Stop()
Exit Function
End If
If DetailSel = "" Then
frmTesters.Grid1.Cols = numCols
frmTesters.Grid1.Rows = 0
frmTesters.Grid1.Col = 0
frmTesters.Grid1.Row = 0
```

Continues

Listing 11.10 *(Continued)*

```
frmTesters.Grid1.ColWidth(0) = 230
Do While (SQLFetch(hstmt) = SQL_SUCCESS)
frmTesters.Grid1.Rows = frmTesters.Grid1.Rows + 1
frmTesters.Grid1.Row = frmTesters.Grid1.Rows - 1
For i = 1 To numCols
rc = SQLGetData(hstmt, i, SQL_C_CHAR, data, 80, retLen)
frmTesters.Grid1.Col = i - 1
frmTesters.Grid1.Text = Trim(data)
If (rc <> 0) Then
rc = DB2Stop()
Exit Function
End If
Next
data = String(40, " ")
Loop
Else
frmTesters.txtTesters.Text = ""
frmTesters.txtTesters.Refresh
Do While (SQLFetch(hstmt) = SQL_SUCCESS)
For i = 1 To numCols
rc = SQLGetData(hstmt, i, SQL_C_CHAR, data, 80, retLen)
frmTesters.txtTesters.Text = frmTesters.txtTesters.Text &
    Trim(data) & vbCrLf
If (rc <> 0) Then
rc = DB2Stop()
Exit Function
End If
frmTesters.txtTesters.Text = frmTesters.txtTesters.Text & vbCrLf
Next
data = String(40, " ")
Loop
DetailSel = ""
End If

rc = DB2Stop()
DB2eTest = 0
End Function
Public Function DB2Stop() As Integer
On Error Resume Next
If rc = 0 Then
frmTesters.txtTesters.Text = frmTesters.txtTesters.Text & "
    Command executed. " & vbCrLf
Else
frmTesters.txtTesters.Text = frmTesters.txtTesters.Text & " Error
    during execution. " & vbCrLf
End If
rc = SQLDisconnect(hdbc)
rc = SQLFreeHandle(SQL_HANDLE_DBC, hdbc)
rc = SQLFreeHandle(SQL_HANDLE_ENV, henv)
DB2Stop = 0
End Function
```

Listing 11.10 Main module for database and table query. *(Continued)*

There is only one more part to do. We have to present the end user with detailed information about the specific tester. To achieve this, tap (click) the desired record in the displayed grid (in our case, record number 3), and the Grid1_Click (lines 26 through 30 in Listing 11.9) event will be executed. In return, the lines 93 through 107 in Listing 11.10 will be executed and the "Status/Details" text area will be populated with desired details. You can see the resulting screen in Figure 11.11.

You should understand that the application and database design are of extremely high importance when you are developing portable applications and mobile databases. Limitations of many kinds exist in the fields of available storage space (most devices on the market have only between 8MB and 50MB of available storage space, without additional cards), display size is much larger than, for example, that of WAP-enabled cell phones, but the density of the data we can "paint" on the display is small compared to desktop applications. In addition, availability of Internet connectivity is limited and intermittent and there are limitations due to battery life cycles. All of the aforementioned limitations influence guidelines for the application development and design. One of the most important is the creation of "drill down" data mining. However, no information should be more than four levels of depth away from the top-level result set; otherwise, the end user's experience with your application is going to be poor, and, in most cases, the user population will not use the application. Database design also plays a major role in application performance and usability. Except in rare cases, portable databases are only subsets of data from the corporate and enterprise databases. As such, try not to replicate unnecessary data to the portable devices, and spend a little bit more time on understanding exact data requirements for your specific application.

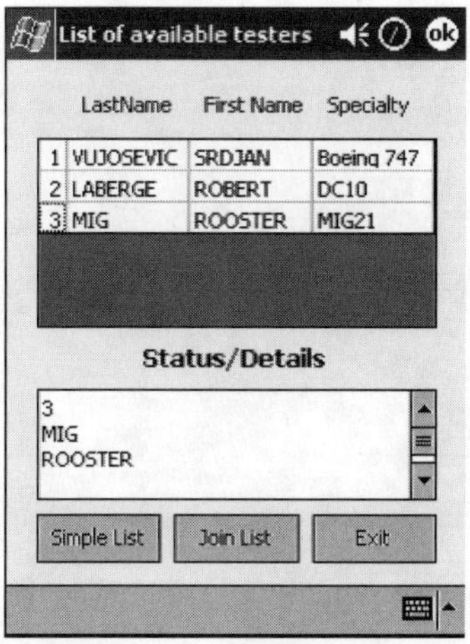

Figure 11.11 Show person details.

Even with all of the limitations that we stated, simple table structures, like the ones presented in this chapter, will allow you to store thousands of data rows.

```
'----------------------------------------------------------------
' File: db2ecli.bas
'
' Description: Visual Basic interface for DB2e database
'
' To do: Add constants from sqlcli.h, sqlcli1.h, sqlext.h,
  sqlsystm.h as needed
'
' Created: July 27, 2001
'
' Version 7.2
'----------------------------------------------------------------
'
' DB2e constants
'
Option Explicit

Const SQL_HANDLE_ENV  As Long = 1
Const SQL_HANDLE_DBC  As Long = 2
Const SQL_HANDLE_STMT As Long = 3
Const SQL_HANDLE_DESC As Long = 4

Const SQL_BINARY            As Long = -2
Const SQL_CHAR              As Long = 1
Const SQL_INTEGER           As Long = 4
Const SQL_SMALLINT          As Long = 5
Const SQL_DATE              As Long = 9
Const SQL_TIME              As Long = 10
Const SQL_TIMESTAMP         As Long = 11
Const SQL_TYPE_DATE         As Long = 91
Const SQL_TYPE_TIME         As Long = 92
Const SQL_TYPE_TIMESTAMP    As Long = 93
Const SQL_C_LONG            As Long = 4
Const SQL_C_SHORT           As Long = 5
Const SQL_C_CHAR            As Long = 1
Const SQL_C_BINARY          As Long = -2
Const SQL_C_DATE            As Long = 9
Const SQL_C_TIME            As Long = 10
Const SQL_C_TIMESTAMP       As Long = 11
Const SQL_C_TYPE_DATE       As Long = 91
Const SQL_C_TYPE_TIME       As Long = 92
Const SQL_C_TYPE_TIMESTAMP  As Long = 93

Const SQL_PARAM_INPUT As Long = 1
Const SQL_CLOSE       As Long = 0
Const SQL_SQLSTATE_SIZE As Long = 5

Const DB2eMajorVesion As Long = 7
```

Listing 11.11 DB2 Everyplace CLI functions module.

```
Const DB2eMinorVesion As Long = 2

Const DB2eOutOfMemory As Long = 57011

'
' Return code value
'
Const SQL_SUCCESS           As Long = 0
Const SQL_SUCCESS_WITH_INFO As Long = 1
Const SQL_NO_DATA_FOUND     As Long = 100
Const SQL_NEED_DATA         As Long = 99
Const SQL_NO_DATA           As Long = 100
Const SQL_STILL_EXECUTING   As Long = 2
Const SQL_ERROR             As Long = -1
Const SQL_INVALID_HANDLE    As Long = -2

Const SQL_NOTHING           As Long = 999

Const NOT_ENOUGH_MEMORY     As Long = -100
Const NOT_VALID_ID          As Long = -101

Const NO_CONVERSIONTABLE         As Long = -200
Const CONVERSIONTABLE_NOTLOADED  As Long = -201

'
'from UDB
'
Const SQL_GRAPHIC           As Long = -95
Const SQL_VARGRAPHIC        As Long = -96
Const SQL_LONGVARGRAPHIC    As Long = -97
Const SQL_BLOB              As Long = -98
Const SQL_CLOB              As Long = -99
Const SQL_DBCLOB            As Long = -350

Const SQL_NULL_DATA         As Long = -1
Const SQL_NTS               As Long = -3

Const MAXVCHAR              As Long = 4000

Const SQL_NULL_HENV As Long = 0
Const SQL_NULL_HDBC As Long = 0
Const SQL_NULL_HSTMT As Long = 0

'
' DB2e functions
'

Declare Function SQLAllocConnect Lib "db2e.dll" Alias
    "DB2eAllocConnect" (ByVal henv As Long, phdbc As Long) As Integer

Declare Function SQLAllocEnv Lib "db2e.dll" Alias "DB2eAllocEnv"
    (phenv As Long) As Integer
```

Continues

Listing 11.11 *(Continued)*

```
82. Declare Function SQLAllocStmt Lib "db2e.dll" Alias "DB2eAllocStmt"
      (ByVal hdbc As Long, phstmt As Long) As Integer

83. Declare Function SQLAllocConnectVer Lib "db2e.dll" Alias
      "DB2eAllocConnectVer" (ByVal henv As Long, phdbc As Long, ByVal
      version As Long) As Integer

84. Declare Function SQLAllocEnvVer Lib "db2e.dll" Alias
      "DB2eAllocEnvVer" (phenv As Long, ByVal version As Long) As Integer

85. Declare Function SQLAllocStmtVer Lib "db2e.dll" Alias
      "DB2eAllocStmtVer" (ByVal hdbc As Long, phstmt As Long,
      ByVal version As Long) As Integer

86. Declare Function SQLAllocHandle Lib "db2e.dll" Alias
      "DB2eAllocHandle" (ByVal HandleType As Integer, ByVal
      InputHandle As Long, OutputHandlePtr As Long) As Integer

87. Declare Function SQLAllocHandleVer Lib "db2e.dll" Alias
      "DB2eAllocHandleVer" ( _
88. ByVal HandleType As Integer, _
89. ByVal InputHandle As Long, _
90. OutputHandlePtr As Long, _
91. ByVal version As Long) _
92. As Integer

93. Declare Function SQLBindCol Lib "db2e.dll" Alias "DB2eBindCol"
      (ByVal hstmt As Long, ByVal icol As Integer, ByVal fCType As
      Integer, ByVal szResult As String, ByVal cbValueMax As Long,
      pcbValue As Long) As Integer

94. Declare Function SQLCancel Lib "db2e.dll" Alias "DB2eCancel"
      (ByVal hstmt As Long) As Integer

95. Declare Function SQLColAttributes Lib "db2e.dll" Alias
      "DB2eColAttributesW" (ByVal hstmt As Long, ByVal icol As
      Integer, ByVal fDescType As Integer, rgbDesc As Variant, ByVal
      cbDescMax As Integer, pcbdesc As Integer, pfdesc As Long) As
      Integer

96. Declare Function SQLConnect Lib "db2e.dll" Alias "DB2eConnectWInt"
      (ByVal hdbc As Long, ByVal szDSN As String, ByVal cbDSN As
      Integer, ByVal szUID As String, ByVal cbUID As Integer, ByVal
      szAuthStr As String, ByVal cbAuthStr As Integer) As Integer

97. Declare Function SQLDescribeCol Lib "db2e.dll" Alias
      "DB2eDescribeColW" (ByVal hstmt As Long, ByVal icol As Integer,
      ByVal szColName As String, ByVal cbColNameMax As Integer,
      pcbColName As Integer, pfSqlType As Integer, pcbColDef As Long,
      pibScale As Integer, pfNullable As Integer) As Integer

98. Declare Function SQLDisconnect Lib "db2e.dll" Alias
      "DB2eDisconnect" (ByVal hdbc As Long) As Integer
```

Listing 11.11 DB2 Everyplace CLI functions module. *(Continued)*

```
Declare Function SQLError Lib "db2e.dll" Alias "DB2eErrorW" (ByVal
    henv As Long, ByVal hdbc As Long, ByVal hstmt As Long, ByVal
    szSqlState As String, pfNativeError As Long, ByVal szErrorMsg
    As String, ByVal cbErrorMsgMax As Integer, pcbErrorMsg As
    Integer) As Integer

Declare Function SQLExecDirect Lib "db2e.dll" Alias
    "DB2eExecDirectW" (ByVal hstmt As Long, ByVal szSqlStr As
    String, ByVal cbSqlStr As Long) As Integer

Declare Function SQLExecute Lib "db2e.dll" Alias "DB2eExecute"
    (ByVal hstmt As Long) As Integer

Declare Function SQLFetch Lib "db2e.dll" Alias "DB2eFetch" (ByVal
    hstmt As Long) As Integer

Declare Function SQLFreeConnect Lib "db2e.dll" Alias
    "DB2eFreeConnect" (ByVal hdbc As Long) As Integer

Declare Function SQLFreeEnv Lib "db2e.dll" Alias "DB2eFreeEnv"
    (ByVal henv As Long) As Integer

Declare Function SQLFreeStmt Lib "db2e.dll" Alias "DB2eFreeStmt"
    (ByVal hstmt As Long, ByVal fOption As Integer) As Integer

Declare Function SQLFreeHandle Lib "db2e.dll" Alias
    "DB2eFreeHandle" (ByVal HandleType As Integer, ByVal Handle As
    Long) As Integer

Declare Function SQLGetCursorName Lib "db2e.dll" Alias
    "DB2eGetCursorNameW" (ByVal hstmt As Long, ByVal szCursor As
    String, ByVal cbCursorMax As Integer, pcbCursor As Integer) As
    Integer

Declare Function SQLGetData Lib "db2e.dll" Alias "DB2eGetData"
    (ByVal hstmt As Long, ByVal icol As Integer, ByVal fCType As
    Integer, ByVal rgbValue As String, ByVal cbValueMax As Long,
    pcbValue As Long) As Integer

Declare Function SQLNumResultCols Lib "db2e.dll" Alias
    "DB2eNumResultCols" (ByVal hstmt As Long, pccol As Integer) As
    Integer

Declare Function SQLPrepare Lib "db2e.dll" Alias "DB2ePrepareW"
    (ByVal hstmt As Long, ByVal szSqlStr As String, ByVal cbSqlStr
    As Long) As Integer

Declare Function SQLRowCount Lib "db2e.dll" Alias "DB2eRowCount"
    (ByVal hstmt As Long, pcrow As Long) As Integer

Declare Function SQLSetCursorName Lib "db2e.dll" Alias
    "DB2eSetCursorNameW" (ByVal hstmt As Long, ByVal szCursor As
    String, ByVal cbCursor As Integer) As Integer
```

Continues

Listing 11.11 *(Continued)*

```
Declare Function SQLSetParam Lib "db2e.dll" Alias "DB2eSetParam" (ByVal hstmt As Long, ByVal ipar As Integer, ByVal fCType As Integer, ByVal fSqlType As Integer, ByVal cbColDef As Long, ByVal ibScale As Integer, rgbValue As Variant, pcbValue As Long) As Integer

Declare Function SQLTransact Lib "db2e.dll" Alias "DB2eTransact" (ByVal henv As Long, ByVal hdbc As Long, ByVal fType As Integer) As Integer

Declare Function SQLEndTran Lib "db2e.dll" Alias "DB2eEndTran" (ByVal HandleType As Integer, ByVal Handle As Long, ByVal CompletionType As Integer) As Integer

Declare Function SQLSetConnectAttr Lib "db2e.dll" Alias "DB2eSetConnectAttrW" (ByVal hdbc As Long, ByVal fOption As Integer, ByVal vParam As Variant, ByVal StrLen As Long) As Integer

Declare Function SQLGetDiagRec Lib "db2e.dll" Alias "DB2eGetDiagRecW" _
(ByVal HandleType As Integer, _
ByVal Handle As Long, _
ByVal RecNumber As Integer, _
ByVal sqlState As String, _
nativeError As Long, _
ByVal messageText As String, _
ByVal BufferLength As Integer, _
textlength As Integer) _
As Integer

Declare Function SQLSetStmtAttr Lib "db2e.dll" Alias "DB2eSetStmtAttrW" (ByVal hstmt As Long, ByVal fOption As Integer, pvParam As Variant, ByVal fStrLen As Integer) As Integer

Declare Function SQLFetchScroll Lib "db2e.dll" (ByVal hstmt As Long, ByVal ori As Integer, ByVal offset As Long) As Integer

Declare Function SQLGetInfo Lib "db2e.dll" Alias "DB2eGetInfoW" (ByVal hdbc As Long, ByVal fInfoType As Integer, ByRef rgbInfoValue As Variant, ByVal cbInfoMax As Integer, cbInfoOut As Integer) As Integer
```

Listing 11.11 DB2 Everyplace CLI functions module. *(Continued)*

Data Synchronization and Replication

The required level of product knowledge dictates that the person installing DB2 Everyplace Sync server must have good experience with IBM DB2 products and a certain level of DB2 synchronization to successfully implement this product. For that reason we will now explain principles of two-way synchronization and show you what processes are involved in synchronization itself.

In order to achieve data synchronization, you will have to install three components: DB2 Everyplace database engine will be installed on the portable device, DB2 Everyplace Sync Server will be installed on the middle-tier system (also known as mirroring server), and DB2 Universal Database is required for the enterprise database. Many mobile devices use one mirroring server to synchronize data. For successful synchronization you must have the same version of the DB2 Everyplace on your portable device and on the Sync Server. The most important part of the whole synchronization process is the IBM sync application that is installed on the portable device together with the DB2 Everyplace database engine.

Before you can start using this application, however, you will have to configure a couple of details. As you can see in Figure 11.12, you will have to configure the IP address of the system where the sync server is loaded, and configure the port number that is used for communication. In addition, you will have to provide the user ID and password of the user who is authorized to perform synchronization. Since DB2 Everyplace synchronization uses TCP/IP connectivity, it is well suited for use over the Internet, as well as for the implementation behind firewalls. The synchronization servlet requires use of a Web server with Java Servlet API 2.0 support. A basic version of WebSphere that satisfies these requirements is included with DB2 Everyplace Enterprise Edition. The image on the right in Figure 11.12 represents a selection of defined synchronization sets that are available to you. Select the set available from the list, and then select the radio button beside the desired synchronization type.

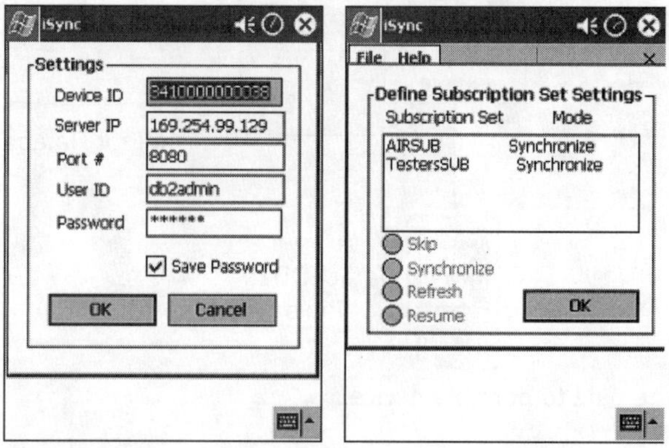

Figure 11.12 Configure data synchronization.

In the following paragraphs, we will introduce you to database synchronization.

Let's show you what the synchronization diagram would be for a corporate database to the portable device. From Figure 11.13 we can immediately identify what the major steps of this process would be:

1. The Source table on the corporate database has been updated.
2. The DB2 DataPropagator Capture program, running continuously on the source system, captures changes to the subscriber database source table from the DB2 log and writes them to the CD table.
3. The DB2 DataPropagator Apply program applies changes from the CD table to the mirror table.
4. Changes to the data are sent to an output queue in the form of a synchronization reply message.
5. The synchronization client (installed on portable device) retrieves the synchronization reply message.

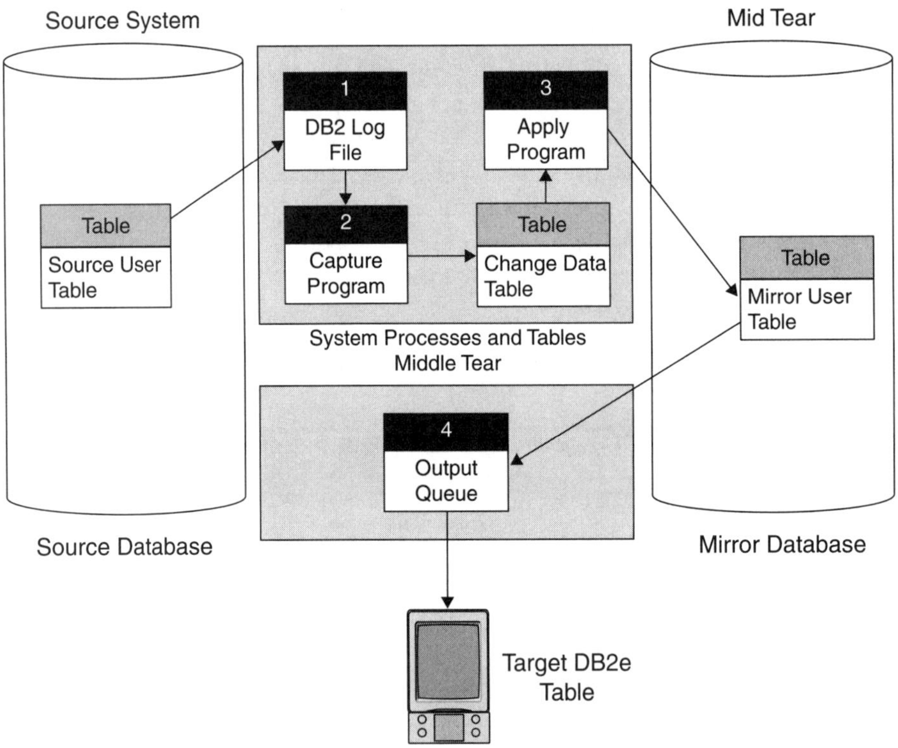

Figure 11.13 Synchronize data to portable device.

6. Changes to the data are applied to the local copy of the table (on the portable device in this case).
7. Synchronization ends.

On the other hand, synchronization from a portable device to the corporate database will follow these steps (please use Figure 11.14 for this reference):

1. The data has been updated on the portable device.
2. The portable device user starts IBM sync application (please refer to Figure 11.14) and selects the name of the synchronization set to execute.
3. The synchronization request is first authenticated and then placed in the Input Queue on the middle-tier server.
4. The user is allowed to perform synchronization of the data subset to which he has been subscribed.
5. The data is placed in the staging table. This design can improve performance in high-volume synchronization environments.
6. The data is copied from staging to the mirror table. At the same time, all of the transactions are logged in DB2 Everyplace log and possible synchronization conflicts are resolved.
7. The capture program detects changes. Changes are written from DB2e log to change data (CD) table.
8. The DB2 Everyplace Apply program starts and propagates changes to CD table to the table on the corporate database.
9. Synchronization ends.

That is the whole process for synchronization, but what will the end users see on their portable devices? Users will be able to see synchronization messages between portable devices and middle-tier servlet applications. It is very important to ensure that the middle-tier Sync Server is installed, configured, and operational at this time. This middle tier will, in the real world, reside between Internet/local network services and DB2 UDB databases. It is not necessary to have any DB2-specific clients installed to check the servlet. The following steps are required for this test:

1. Start the Web application server (by selecting appropriate options in the Start menu).
2. In a Web browser, go to http://*(your middle tier system)*: 8080/db2e/db2erdb.

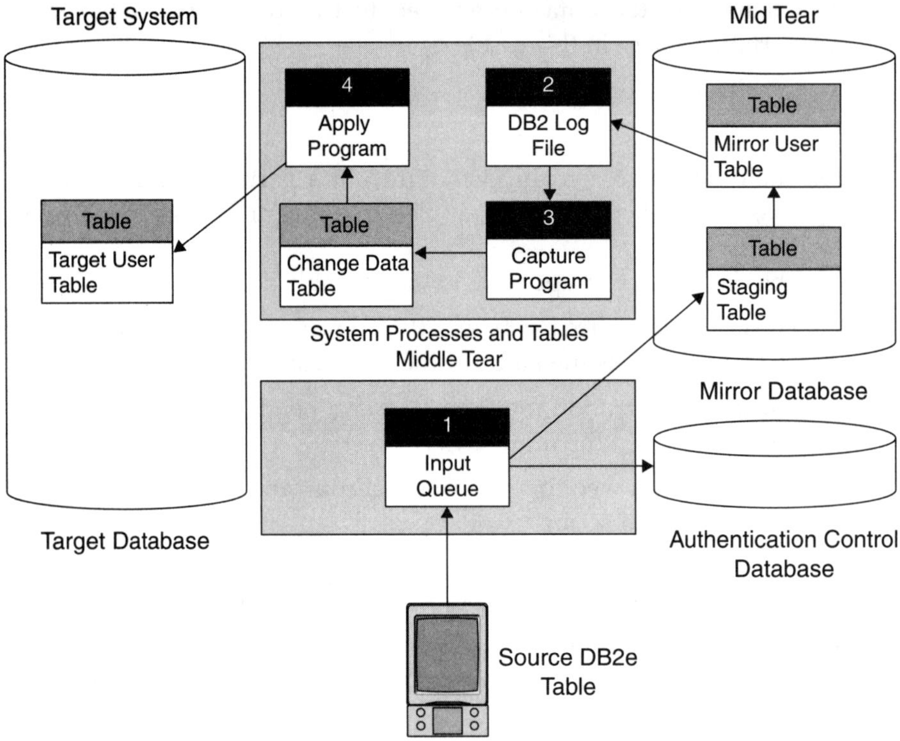

Figure 11.14 Synchronize data from portable device.

3. You should receive the following message:

```
com.ibm.mobileservices.adapter.rdb.RDb <date> <time>
DB2 Everyplace SyncServer 7.2.1 <build date>
```

In case port 8080 has been closed and you have to change the processes, you have to locate the following file: $DSYINSTDIR/WebSphere/AppSrver/bin /servletengine.bat. Near the bottom of the entry list, replace xxxx with the number that is available.

```
Dcontrolserver.websphere.websphere.port=xxxx ...
```

Be careful when selecting the ports because you can affect the system and system performance if you pick one of the ports that are used by the system.

When the process is completed, the end user can review the synchronization log as you can see in Figure 11.15. This technology (synchronization) will become one of the drivers for the success of the overall database engine. Ease

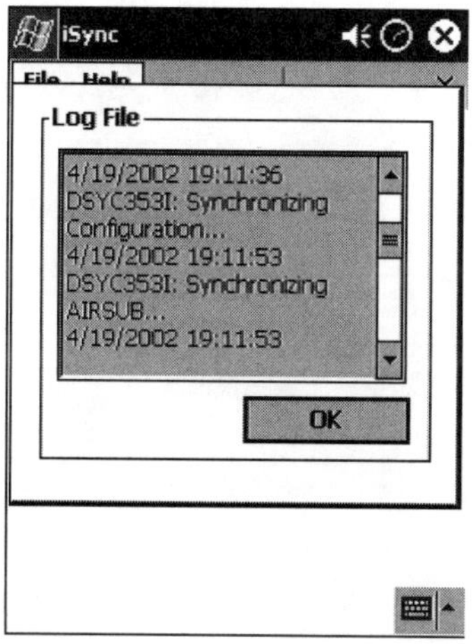

Figure 11.15 Synchronization.

of use and setup of synchronization will play a crucial role in database selection criteria.

Happy programming with DB2 Everyplace.

Final Thoughts

IBM is a well-rounded company that offers coverage for all major platforms and operating systems. To set up and maintain synchronization, the initial installation will not be as simple as Microsoft's. Often you will be interrupted and required to execute parts of the process from the command prompt. However, at the end of this process is a very stable product that scales from Palm-based devices to mainframes.

APPENDIX A

Palm Conduits

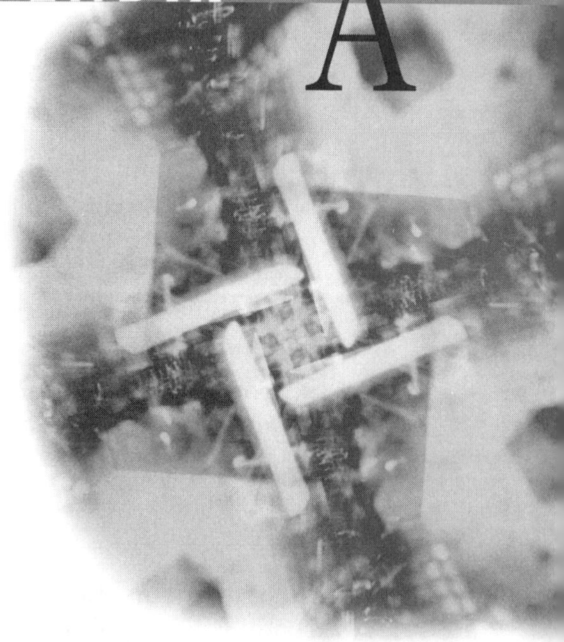

This appendix explains the basic principles of conduits, what they are, the different types of conduits that exist, and basic steps to building conduits with COM objects.

The basic definition of a conduit is an add-on application to the HotSync Manager that will manage the transfer of data from the desktop to the Palm device, and vice versa. At the same time, depending on the type of conduit, it will manage data conversion to the appropriate format.

Conduits are not specific to databases, but rather to each application being Hotsync'd. It's the connectivity software that determines how records are handled during synchronization. Each application has its own conduit. It's the connectivity middleware per application (mail, memo, to-do, address list, and so on) encompassing the specific details and parameters that tells Hotsync how to synchronize each record between the Palm handheld device and another software (such as the Palm Desktop software).

To satisfy the basic functionality, conduits must meet the following requirements:

- Open and close databases, and read records on the Palm powered device.
- Add, delete, and modify records on the Palm device and desktop computer, and simultaneously convert data to appropriate formats.

- Be able to compare records based on FastSync specification (read more about this at http://oasis.palm.com/dev/kb/manuals/knowledgebasearticle.cfm?article_id=1684#925394).

The type of synchronization performed can divide conduits into three basic categories:

- Conduits that synchronize between custom databases on both ends of the conduit (PDA and PC) (Figure A.1).
- Conduits that synchronize between the custom database on the PC side and the standard application database on the PDA side (Figure A.2).
- Conduits that synchronize between standard application databases on the PDA and the PC side (Figure A.3).

What is required to create a conduit, and what is the process? To design new conduits, you will have to first download CDK 4.0 by following www.palmos.com/dev/tech/conduits/. This download will enable you to create new conduits using your favorite development language, Visual Basic, C/C++, or Java. Obviously, you will need C/C++, Visual Basic, or Java IDE/compiler available to complete the list of requirements.

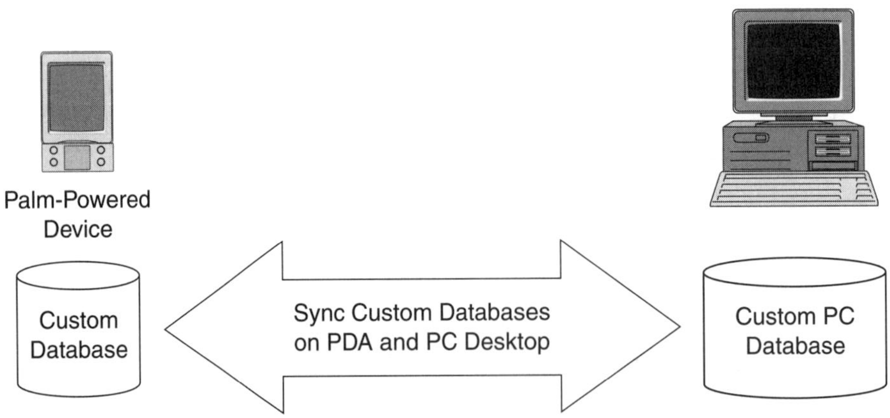

Figure A.1 Synchronization between custom databases on both ends of the conduit.

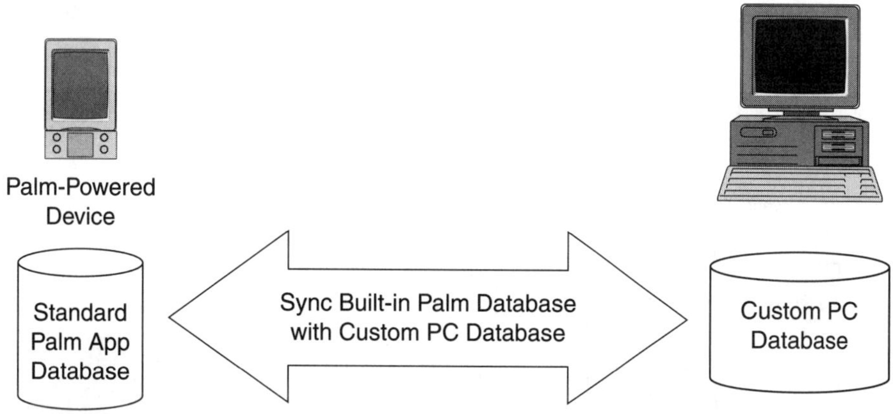

Figure A.2 Synchronization between the custom PC database and the standard PDA application database.

What goals should you follow during the design process? Generally, you should follow these major points:

- Fast execution
- No data loss
- Good exception and conflict handling
- No user interaction

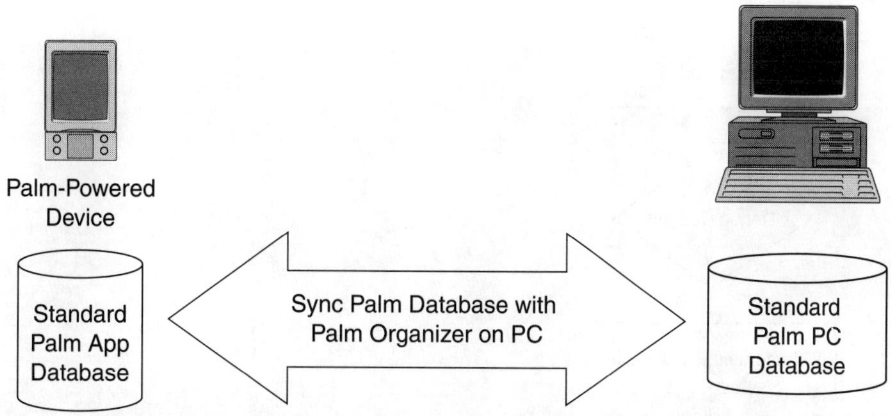

Figure A.3 Synchronization between standard PDA and PC application databases.

Figure A.4 is a flow chart that we recommend you follow if you are going to develop your own conduit.

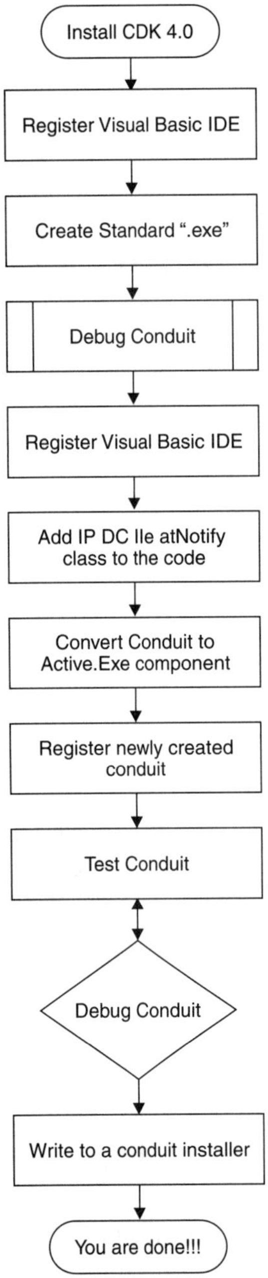

Figure A.4 Process flow in conduit creation process.

APPENDIX B

Microsoft Publication Wizard Script

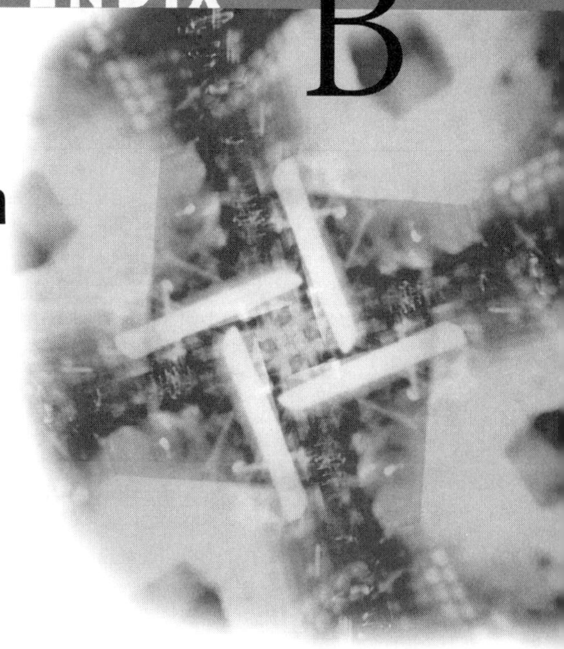

```
/****** Scripting the replication setup of server {Your Server Name}.
    Script Date: 3/25/2002 12:18:55 AM ******/
/****** Please note: Any password parameter was scripted with NULL or
    empty string for security reason. ******/

/****** Begin: Script to be run at Distributor: {Your Server Name} ******/
/****** Object:  Job Reinitialize subscriptions having data validation
    failures    Script Date: 3/25/2002 12:19:00 AM ******/
begin transaction
  DECLARE @JobID BINARY(16)
  DECLARE @ReturnCode INT
  SELECT @ReturnCode = 0
if (select count(*) from msdb.dbo.syscategories where name =
    N'REPL-Alert Response') < 1
  execute msdb.dbo.sp_add_category N'REPL-Alert Response'

select @JobID = job_id from msdb.dbo.sysjobs where (name = N'Reinitial-
    ize subscriptions having data validation failures')
if (@JobID is NULL)
BEGIN
  execute @ReturnCode = msdb.dbo.sp_add_job @job_id = @JobID OUTPUT,
    @job_name = N'Reinitialize subscriptions having data validation
    failures', @enabled = 1, @start_step_id = 1, @notify_level_eventlog
    = 0, @notify_level_email = 0, @notify_level_netsend = 0,
    @notify_level_page = 0, @delete_level = 0, @description =
    N'Reinitializes all subscriptions that have data validation fail-
    ures.', @category_name = N'REPL-Alert Response', @owner_login_name =
    N'{Your Server Name}\Administrator'
  if (@@ERROR <> 0 OR @ReturnCode <> 0) goto QuitWithRollback
```

```
  execute @ReturnCode = msdb.dbo.sp_add_jobstep @job_id = @JobID ,
    @step_id = 1, @cmdexec_success_code = 0, @on_success_action = 1,
    @on_success_step_id = 0, @on_fail_action = 2, @on_fail_step_id = 0,
    @retry_attempts = 0, @retry_interval = 0, @os_run_priority = 0,
    @flags = 0, @step_name = N'Run agent.', @subsystem = N'TSQL',
    @command = N'exec dbo.sp_MSreinit_failed_subscriptions
    @failure_level = 1', @server = N'{Your Server Name}', @database_name
    = N'master'
  if (@@ERROR <> 0 OR @ReturnCode <> 0) goto QuitWithRollback

  execute @ReturnCode = msdb.dbo.sp_update_job @job_id = @JobID,
    @start_step_id = 1
  if (@@ERROR <> 0 OR @ReturnCode <> 0) goto QuitWithRollback

  execute @ReturnCode = msdb.dbo.sp_add_jobserver @job_id = @JobID,
    @server_name = N'{Your Server Name}'
  if (@@ERROR <> 0 OR @ReturnCode <> 0) goto QuitWithRollback


END

commit transaction
goto EndSave
QuitWithRollback:
  if (@@TRANCOUNT > 0) rollback transaction
EndSave:

GO

/****** Object:  Job Replication agents checkup    Script Date:
    3/25/2002 12:19:01 AM ******/
begin transaction
  DECLARE @JobID BINARY(16)
  DECLARE @ReturnCode INT
  SELECT @ReturnCode = 0
if (select count(*) from msdb.dbo.syscategories where name =
    N'REPL-Checkup') < 1
  execute msdb.dbo.sp_add_category N'REPL-Checkup'

select @JobID = job_id from msdb.dbo.sysjobs where (name = N'Replication
    agents checkup')
if (@JobID is NULL)
BEGIN
  execute @ReturnCode = msdb.dbo.sp_add_job @job_id = @JobID OUTPUT,
    @job_name = N'Replication agents checkup', @enabled = 1,
    @start_step_id = 1, @notify_level_eventlog = 2, @notify_level_email
    = 0, @notify_level_netsend = 0, @notify_level_page = 0,
    @delete_level = 0, @description = N'Detects replication agents that
    are not logging history actively.', @category_name = N'REPL-
    Checkup', @owner_login_name = N'{Your Server Name}\Administrator'
  if (@@ERROR <> 0 OR @ReturnCode <> 0) goto QuitWithRollback
```

```
    execute @ReturnCode = msdb.dbo.sp_add_jobstep @job_id = @JobID ,
      @step_id = 1, @cmdexec_success_code = 0, @on_success_action = 1,
      @on_success_step_id = 0, @on_fail_action = 2, @on_fail_step_id = 0,
      @retry_attempts = 0, @retry_interval = 0, @os_run_priority = 0,
      @flags = 0, @step_name = N'Run agent.', @subsystem = N'TSQL',
      @command = N'sp_replication_agent_checkup @heartbeat_interval = 10',
      @database_name = N'master'
    if (@@ERROR <> 0 OR @ReturnCode <> 0) goto QuitWithRollback

    execute @ReturnCode = msdb.dbo.sp_update_job @job_id = @JobID,
      @start_step_id = 1
    if (@@ERROR <> 0 OR @ReturnCode <> 0) goto QuitWithRollback

    execute @ReturnCode = msdb.dbo.sp_add_jobschedule @job_id = @JobID,
      @name = N'Replication agent schedule.', @enabled = 1, @freq_type =
      4, @freq_interval = 1, @freq_subday_type = 4, @freq_subday_interval
      = 10, @freq_relative_interval = 1, @freq_recurrence_factor = 0,
      @active_start_date = 20020301, @active_end_date = 99991231,
      @active_start_time = 0, @active_end_time = 235959
    if (@@ERROR <> 0 OR @ReturnCode <> 0) goto QuitWithRollback

    execute @ReturnCode = msdb.dbo.sp_add_jobserver @job_id = @JobID,
      @server_name = N'{Your Server Name}'
    if (@@ERROR <> 0 OR @ReturnCode <> 0) goto QuitWithRollback


END

commit transaction
goto EndSave
QuitWithRollback:
  if (@@TRANCOUNT > 0) rollback transaction
EndSave:

GO

/****** Installing the server {Your Server Name} as a Distributor.
    Script Date: 3/25/2002 12:19:01 AM ******/
use master
GO

exec sp_adddistributor  @distributor = N'{Your Server Name}', @password
    = N'', @from_scripting = 1
GO

-- Updating the agent profile defaults
sp_MSupdate_agenttype_default @profile_id = 1
GO
```

```
sp_MSupdate_agenttype_default @profile_id = 2
GO
sp_MSupdate_agenttype_default @profile_id = 4
GO
sp_MSupdate_agenttype_default @profile_id = 6
GO
sp_MSupdate_agenttype_default @profile_id = 11
GO

-- Adding the distribution database
exec sp_adddistributiondb  @database = N'distribution', @data_folder =
    N'C:\Program Files\Microsoft SQL Server\MSSQL\Data', @data_file =
    N'distribution.MDF', @data_file_size = 2, @log_folder = N'C:\Program
    Files\Microsoft SQL Server\MSSQL\Data', @log_file =
    N'distribution.LDF', @log_file_size = 0, @min_distretention = 0,
    @max_distretention = 72, @history_retention = 48, @security_mode =
    1, @from_scripting = 1
GO

/****** Object:  Job Distribution clean up: distribution    Script Date:
    3/25/2002 12:19:01 AM ******/
begin transaction
  DECLARE @JobID BINARY(16)
  DECLARE @ReturnCode INT
  SELECT @ReturnCode = 0
if (select count(*) from msdb.dbo.syscategories where name =
    N'REPL-Distribution Cleanup') < 1
  execute msdb.dbo.sp_add_category N'REPL-Distribution Cleanup'

select @JobID = job_id from msdb.dbo.sysjobs where (name =
    N'Distribution clean up: distribution')
if (@JobID is NULL)
BEGIN
  execute @ReturnCode = msdb.dbo.sp_add_job @job_id = @JobID OUTPUT,
    @job_name = N'Distribution clean up: distribution', @enabled = 0,
    @start_step_id = 1, @notify_level_eventlog = 0, @notify_level_email
    = 0, @notify_level_netsend = 0, @notify_level_page = 0,
    @delete_level = 0, @description = N'Removes replicated transactions
    from the distribution database.', @category_name =
    N'REPL-Distribution Cleanup', @owner_login_name = N'{Your Server
    Name}\Administrator'
  if (@@ERROR <> 0 OR @ReturnCode <> 0) goto QuitWithRollback


  execute @ReturnCode = msdb.dbo.sp_add_jobstep @job_id = @JobID ,
    @step_id = 1, @cmdexec_success_code = 0, @on_success_action = 1,
    @on_success_step_id = 0, @on_fail_action = 2, @on_fail_step_id = 0,
    @retry_attempts = 0, @retry_interval = 0, @os_run_priority = 0,
    @flags = 0, @step_name = N'Run agent.', @subsystem = N'TSQL',
    @command = N'EXEC dbo.sp_MSdistribution_cleanup @min_distretention =
    0, @max_distretention = 72', @server = N'{Your Server Name}',
```

```
    @database_name = N'distribution'
  if (@@ERROR <> 0 OR @ReturnCode <> 0) goto QuitWithRollback

  execute @ReturnCode = msdb.dbo.sp_update_job @job_id = @JobID,
    @start_step_id = 1
  if (@@ERROR <> 0 OR @ReturnCode <> 0) goto QuitWithRollback

  execute @ReturnCode = msdb.dbo.sp_add_jobschedule @job_id = @JobID,
    @name = N'Replication agent schedule.', @enabled = 1, @freq_type =
    4, @freq_interval = 1, @freq_subday_type = 4, @freq_subday_interval
    = 10, @freq_relative_interval = 1, @freq_recurrence_factor = 0,
    @active_start_date = 20020301, @active_end_date = 99991231,
    @active_start_time = 500, @active_end_time = 459
  if (@@ERROR <> 0 OR @ReturnCode <> 0) goto QuitWithRollback

  execute @ReturnCode = msdb.dbo.sp_add_jobserver @job_id = @JobID,
    @server_name = N'{Your Server Name}'
  if (@@ERROR <> 0 OR @ReturnCode <> 0) goto QuitWithRollback


END

commit transaction
goto EndSave
QuitWithRollback:
  if (@@TRANCOUNT > 0) rollback transaction
EndSave:

GO

/****** Object:  Job Agent history clean up: distribution    Script
    Date: 3/25/2002 12:19:01 AM ******/
begin transaction
  DECLARE @JobID BINARY(16)
  DECLARE @ReturnCode INT
  SELECT @ReturnCode = 0
if (select count(*) from msdb.dbo.syscategories where name =
    N'REPL-History Cleanup') < 1
  execute msdb.dbo.sp_add_category N'REPL-History Cleanup'

select @JobID = job_id from msdb.dbo.sysjobs where (name =
    N'Agent history clean up: distribution')
if (@JobID is NULL)
BEGIN
  execute @ReturnCode = msdb.dbo.sp_add_job @job_id = @JobID OUTPUT,
    @job_name = N'Agent history clean up: distribution', @enabled = 1,
    @start_step_id = 1, @notify_level_eventlog = 0, @notify_level_email
    = 0, @notify_level_netsend = 0, @notify_level_page = 0,
    @delete_level = 0, @description = N'Removes replication agent
    history from the distribution database.', @category_name =
    N'REPL-History Cleanup', @owner_login_name = N'{Your Server
```

```
    Name}\Administrator'
  if (@@ERROR <> 0 OR @ReturnCode <> 0) goto QuitWithRollback


  execute @ReturnCode = msdb.dbo.sp_add_jobstep @job_id = @JobID ,
    @step_id = 1, @cmdexec_success_code = 0, @on_success_action = 1,
    @on_success_step_id = 0, @on_fail_action = 2, @on_fail_step_id = 0,
    @retry_attempts = 0, @retry_interval = 0, @os_run_priority = 0,
    @flags = 0, @step_name = N'Run agent.', @subsystem = N'TSQL',
    @command = N'EXEC dbo.sp_MShistory_cleanup @history_retention = 48',
    @server = N'{Your Server Name}', @database_name = N'distribution'
  if (@@ERROR <> 0 OR @ReturnCode <> 0) goto QuitWithRollback

  execute @ReturnCode = msdb.dbo.sp_update_job @job_id = @JobID,
    @start_step_id = 1
  if (@@ERROR <> 0 OR @ReturnCode <> 0) goto QuitWithRollback

  execute @ReturnCode = msdb.dbo.sp_add_jobschedule @job_id = @JobID,
    @name = N'Replication agent schedule.', @enabled = 1, @freq_type =
    4, @freq_interval = 1, @freq_subday_type = 4, @freq_subday_interval
    = 10, @freq_relative_interval = 1, @freq_recurrence_factor = 0,
    @active_start_date = 20020301, @active_end_date = 99991231,
    @active_start_time = 0, @active_end_time = 235959
  if (@@ERROR <> 0 OR @ReturnCode <> 0) goto QuitWithRollback

  execute @ReturnCode = msdb.dbo.sp_add_jobserver @job_id = @JobID,
    @server_name = N'{Your Server Name}'
  if (@@ERROR <> 0 OR @ReturnCode <> 0) goto QuitWithRollback


END

commit transaction
goto EndSave
QuitWithRollback:
  if (@@TRANCOUNT > 0) rollback transaction
EndSave:

GO

-- Adding the distribution publisher
exec sp_adddistpublisher  @publisher = N'{Your Server Name}',
    @distribution_db = N'distribution', @security_mode = 1,
    @working_directory = N'\\{Your Server Name}\CEREPL', @trusted =
    N'false', @thirdparty_flag = 0
GO

/****** Object:  Job {Your Server Name}-Survey-chapter-10    Script
    Date: 3/25/2002 12:19:01 AM ******/
begin transaction
  DECLARE @JobID BINARY(16)
```

```
  DECLARE @ReturnCode INT
  SELECT @ReturnCode = 0
if (select count(*) from msdb.dbo.syscategories where name =
    N'REPL-Snapshot') < 1
  execute msdb.dbo.sp_add_category N'REPL-Snapshot'

select @JobID = job_id from msdb.dbo.sysjobs where (name = N'{Your
    Server Name}-Survey-chapter-10')
if (@JobID is NULL)
BEGIN
  execute @ReturnCode = msdb.dbo.sp_add_job @job_id = @JobID OUTPUT,
    @job_name = N'{Your Server Name}-Survey-chapter-10', @enabled = 1,
    @start_step_id = 1, @notify_level_eventlog = 0, @notify_level_email
    = 0, @notify_level_netsend = 0, @notify_level_page = 0,
    @delete_level = 0, @description = N'No description available.',
    @category_name = N'REPL-Snapshot', @owner_login_name = N'{Your
    Server Name}\Administrator'
  if (@@ERROR <> 0 OR @ReturnCode <> 0) goto QuitWithRollback


  execute @ReturnCode = msdb.dbo.sp_add_jobstep @job_id = @JobID ,
    @step_id = 1, @cmdexec_success_code = 0, @on_success_action = 3,
    @on_success_step_id = 0, @on_fail_action = 3, @on_fail_step_id = 0,
    @retry_attempts = 0, @retry_interval = 0, @os_run_priority = 0,
    @flags = 0, @step_name = N'Snapshot Agent startup message.',
    @subsystem = N'TSQL', @command = N'sp_MSadd_snapshot_history
    @perfmon_increment = 0,  @agent_id = 10, @runstatus = 1,
                    @comments = ''Starting agent.''', @server = N'{Your
    Server Name}', @database_name = N'distribution'
  if (@@ERROR <> 0 OR @ReturnCode <> 0) goto QuitWithRollback

  execute @ReturnCode = msdb.dbo.sp_add_jobstep @job_id = @JobID ,
    @step_id = 2, @cmdexec_success_code = 0, @on_success_action = 1,
    @on_success_step_id = 0, @on_fail_action = 3, @on_fail_step_id = 0,
    @retry_attempts = 10, @retry_interval = 1, @os_run_priority = 0,
    @flags = 0, @step_name = N'Run agent.', @subsystem = N'Snapshot',
    @command = N'-Publisher [{Your Server Name}] -PublisherDB [Survey] -
    Distributor [{Your Server Name}] -Publication [chapter] -Replica-
    tionType 2 -DistributorSecurityMode 1 ', @server = N'{Your Server
    Name}', @database_name = N'distribution'
  if (@@ERROR <> 0 OR @ReturnCode <> 0) goto QuitWithRollback

  execute @ReturnCode = msdb.dbo.sp_add_jobstep @job_id = @JobID ,
    @step_id = 3, @cmdexec_success_code = 0, @on_success_action = 2,
    @on_success_step_id = 0, @on_fail_action = 2, @on_fail_step_id = 0,
    @retry_attempts = 0, @retry_interval = 0, @os_run_priority = 0,
    @flags = 0, @step_name = N'Detect nonlogged agent shutdown.',
    @subsystem = N'TSQL', @command = N'sp_MSdetect_nonlogged_shutdown
    @subsystem = ''Snapshot'', @agent_id = 10', @server = N'{Your Server
    Name}', @database_name = N'distribution'
  if (@@ERROR <> 0 OR @ReturnCode <> 0) goto QuitWithRollback
```

```
  execute @ReturnCode = msdb.dbo.sp_update_job @job_id = @JobID,
    @start_step_id = 1
  if (@@ERROR <> 0 OR @ReturnCode <> 0) goto QuitWithRollback

  execute @ReturnCode = msdb.dbo.sp_add_jobschedule @job_id = @JobID,
    @name = N'Replication agent schedule.', @enabled = 1, @freq_type =
    8, @freq_interval = 64, @freq_subday_type = 1, @freq_subday_interval
    = 0, @freq_relative_interval = 0, @freq_recurrence_factor = 1,
    @active_start_date = 20020308, @active_end_date = 99991231,
    @active_start_time = 231300, @active_end_time = 235959
  if (@@ERROR <> 0 OR @ReturnCode <> 0) goto QuitWithRollback

  execute @ReturnCode = msdb.dbo.sp_add_jobserver @job_id = @JobID,
    @server_name = N'{Your Server Name}'
  if (@@ERROR <> 0 OR @ReturnCode <> 0) goto QuitWithRollback


END

commit transaction
goto EndSave
QuitWithRollback:
  if (@@TRANCOUNT > 0) rollback transaction
EndSave:

GO


/****** End: Script to be run at Distributor: {Your Server Name} ******/

-- Enabling the replication database
use master
GO

exec sp_replicationdboption @dbname = N'Survey', @optname = N'merge pub-
    lish', @value = N'true'
GO

use [Survey]
GO

-- Adding the merge publication
exec sp_addmergepublication @publication = N'chapter', @description =
    N'Merge publication of Survey database from Publisher {Your Server
    Name}.', @retention = 14, @sync_mode = N'character', @allow_push =
    N'true', @allow_pull = N'true', @allow_anonymous = N'true',
    @enabled_for_internet = N'false', @centralized_conflicts = N'true',
    @dynamic_filters = N'false', @snapshot_in_defaultfolder = N'true',
    @compress_snapshot = N'false', @ftp_port = 21, @ftp_login =
    N'anonymous', @conflict_retention = 14, @keep_partition_changes =
```

```
    N'false', @allow_subscription_copy = N'false',
    @allow_synctoalternate = N'false', @add_to_active_directory =
    N'false', @max_concurrent_merge = 0, @max_concurrent_dynamic_
    snapshots = 0
exec sp_addpublication_snapshot @publication =
    N'chapter',@frequency_type = 4, @frequency_interval = 1,
    @frequency_relative_interval = 1, @frequency_recurrence_factor = 0,
    @frequency_subday = 1, @frequency_subday_interval = 5,
    @active_start_date = 0, @active_end_date = 0,
    @active_start_time_of_day = 500, @active_end_time_of_day = 235959,
    @snapshot_job_name = N'{Your Server Name}-Survey-chapter-10'
GO

exec sp_grant_publication_access @publication = N'chapter', @login =
    N'{Your Server Name}\Administrator'
GO
exec sp_grant_publication_access @publication = N'chapter', @login =
    N'{Your Server Name}\{Your IUSR_ComputerName}'
GO
exec sp_grant_publication_access @publication = N'chapter', @login =
    N'BUILTIN\Administrators'
GO
exec sp_grant_publication_access @publication = N'chapter', @login =
    N'distributor_admin'
GO
exec sp_grant_publication_access @publication = N'chapter', @login =
    N'sa'
GO

-- Adding the merge articles
exec sp_addmergearticle @publication = N'chapter', @article =
    N'tblSurveyResults', @source_owner = N'dbo', @source_object =
    N'tblSurveyResults', @type = N'table', @description = null,
    @column_tracking = N'true', @pre_creation_cmd = N'drop',
    @creation_script = null, @schema_option = 0x000000000000CFF1,
    @article_resolver = null, @subset_filterclause =
    N'tblSurveyResults.Upload = ''Y''', @vertical_partition = N'false',
    @destination_owner = N'dbo', @auto_identity_range = N'false',
    @verify_resolver_signature = 0, @allow_interactive_resolver =
    N'false', @fast_multicol_updateproc = N'true', @check_permissions = 0
GO
exec sp_addmergearticle @publication = N'chapter', @article =
    N'tblSurveys', @source_owner = N'dbo', @source_object = N'tblSur-
    veys', @type = N'table', @description = null, @column_tracking =
    N'true', @pre_creation_cmd = N'drop', @creation_script = null,
    @schema_option = 0x000000000000CFF1, @article_resolver = null,
    @subset_filterclause = null, @vertical_partition = N'false',
    @destination_owner = N'dbo', @auto_identity_range = N'false', @ver-
    ify_resolver_signature = 0, @allow_interactive_resolver = N'false',
    @fast_multicol_updateproc = N'true', @check_permissions = 0
GO
```

APPENDIX C

DB2 CLI/ODBC Functions

CLI/ODBC FUNCTIONS RELATED TO DATABASE CONNECTION	
SQLAllocHandle	Obtains a handle.
SQLAllocConnect	Obtains a connection handle.
SQLAllocEnv	Obtains an environment handle.
SQLConnect	Connects to a specific driver by specifying DSN, user ID, and password.
SQLSetConnectAttr	Sets connection-related attributes.

PREPARATION FOR SQL REQUEST	
SQLAllocStmt	Allocates a statement handle.
SQLBindParameter	Allocates storage for the parameters in an SQL statement. The parameter type is limited to INPUT because stored procedures are not supported.
SQLPrepare	Prepares an SQL statement for subsequent execution.
SQLSetStmtAttr	Sets options related to a statement.

SUBMITTING REQUESTS	
SQLExecDirect	Executes a statement. Only Synchronous calls are supported.
SQLExecute	Executes a prepared statement. All parameters must be bound before calling SQLExecute(). Only Synchronous calls are supported.
SQLEndTran	Requests a COMMIT or ROLLBACK for all operations on all statements open with a connection.

RETRIEVING RESULTS AND RESULT STATUS	
SQLBindCol	Allocates storage for a result column and specifies the data type.
SQLDescribeCol	Describes a column in the result set. The column information is limited by the supported column data types.
SQLError	Returns additional error information.
SQLFetch	Returns a result row. The result is retrieved one row at a time and not as result sets.
SQLFetchScroll	Returns a resulting row set. The result is fetched by result sets.
SQLGetData	Returns part or all of one column of one row of a result set.
SQLGetDiagRec	Gets multiple fields of diagnostic data. Only diagnostic records associated with a statement handle are supported.
SQLGetInfo	Provides the DB2 Everyplace version.
SQLNumResultCols	Returns the number of columns in the result set.
SQLRowCount	Returns the number of rows affected by an insert, update, or delete SQL statement.

STOPPING A STATEMENT	
SQLFreeHandle	Frees handle resources.
SQLFreeStmt	Ends statement processing, dismisses pending results.

TERMINATING A CONNECTION	
SQLDisconnect	Closes the connection.
SQLFreeConnect	Releases the connection handle.
SQLFreeEnv	Releases the environment handle.

APPENDIX D

Sybase Glossary

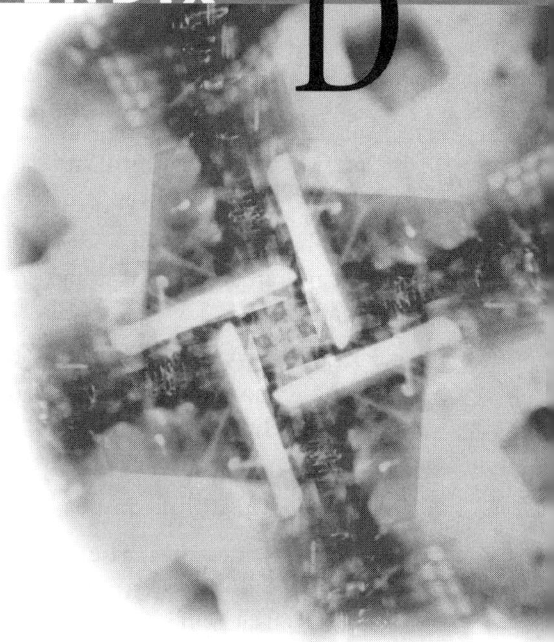

client In MobiLink contexts, client can refer to any application, database engine, or executable that receives data resources from a server or requests a service.

client communication stream Clients can communicate with the synchronization server by way of a number of supported communications protocols.

consolidated database A database that contains all of the data; typically, an enterprise-level database. Supported products include Oracle, IBM's DB2, Microsoft SQL Server, Adaptive Server Anywhere, and Adaptive Server Enterprise.

download That stage in synchronization during which data is transferred from a consolidated database to a remote database.

MobiLink client There are two kinds of MobiLink clients. For Adaptive Server Anywhere remote databases, the MobiLink client is the dbmlsync command-line utility. For UltraLite remote databases, the MobiLink client is built into the UltraLite runtime library.

MobiLink synchronization server A Sybase session-based synchronization technology designed to synchronize UltraLite and Adaptive Server Anywhere databases with industry-standard SQL database-management systems.

publication A database object on the remote database that identifies data to be synchronized. A publication consists of articles that identify tables and columns to be synchronized.

redirector A Web server plug-in that routes requests and responses between a client and the MobiLink synchronization server. This plug-in also implements load-balancing and fail-over mechanisms.

remote database An Adaptive Server Anywhere or UltraLite database that exchanges synchronization messages with a consolidated database. Remote databases can share all or some of the data in the consolidated database.

scripts Written for MobiLink events, scripts programmatically control data exchange to meet business needs.

session-based synchronization A type of synchronization where a synchronization results in consistent data representation across both the consolidated and remote databases.

subscription A database object that serves as a link in a remote database between a publication and a MobiLink user, allowing the data described by the publication to be synchronized.

synchronization Applied to databases, synchronization is the coordination of data between multiple databases for consistent data representation.

transactional integrity The guaranteed maintenance of transactions across the synchronization system. Either a complete transaction is synchronized, or no part of the transaction is synchronized.

upload That stage in synchronization during which data is transferred from a remote database to a consolidated database.

Acronyms

ADOCE ActiveX Data Objects for CE

BLOB Binary Large OBject

BOF Beginning Of File

CDbK Conduit Development Kit

CF Compact Flash

CLI Call Level Interface

DBA DataBase Administrator

DBCS Double Character Sets

DDL Data Definition Language

DLL Dynamic Link Library

DML Data Manipulation Language

DSN Data Source Name

EOF End Of File

ERP Enterprise Resource Planning

eVB Embedded Visual Basic

GB Giga Byte

GUI Graphical User Interface

HPC Handheld Personal Computers

IBM International Business Machines

IDE Integrated Development Environment

IIS Microsoft Internet Information Services

IR Infra Red

JDBC Java DataBase Connectivity

LAN Local Area Network

MDAC Microsoft Data Access Components

MMC Multi Media Card

MSMQ MicroSoft Message Queue

ODBC Open Database Connectivity

OS Operating System

PC Personal Computer

PDA Personal Digital Assistant

PDB Palm Database, used to transfer data to Palm devices

POSE Palm OS Emulator

PQA Palm Query Application

PRC Palm Resource

RDA Remote Data Access

RDBMS Relational Database Management System

ROWGUIDCOL Row Global Unique IDentifier COLumn

RS Record Set

SBCS Single Byte Character Sets

SQL Structured Query Language

SSL Secure Sockets Layer

UCD Universal Conduit Descriptor

UDB Universal DataBase

VB Visual Basic

WAN Wide Area Network

WYSIWYG What You See Is What You Get

XML eXtensible Markup Language

INDEX